ASIC SYSTEM DESIGN WITH VHDL:
A PARADIGM

THE KLUWER INTERNATIONAL SERIES
IN ENGINEERING AND COMPUTER SCIENCE

VLSI, COMPUTER ARCHITECTURE AND
DIGITAL SIGNAL PROCESSING

Consulting Editor
Jonathan Allen

ASIC SYSTEM DESIGN WITH VHDL: A PARADIGM

by

Steven S. Leung
Michigan State University

and

Michael A. Shanblatt
Michigan State University

KLUWER ADADEMIC PUBLISHERS
Boston/Dordrecht/London

Distributors for North America:
Kluwer Academic Publishers
101 Philip Drive
Assinippi Park
Norwell, Massachusetts 02061, USA

Distributors for all other countries:
Kluwer Academic Publishers Group
Distribution Centre
Post Office Box 322
3300 AH Dordrecht, THE NETHERLANDS

Library of Congress Cataloging-in-Publication Data

Leung, Steven S., 1954–
  ASIC system design with VHDL.

  (The Kluwer international series in engineering and
computer science. VLSI, computer architecture, and
digital signal processing)
  Bibliography: p.
  Includes index.
  1. Integrated circuits—Very large scale integration—
Design and construction—Data processing.   2. VHDL
(Computer program language)   I. Shanblatt, Michael A.,
1952–   . II. Title.   III. Series.
TK7874.L396   1989   621.39′5′0285   89-11091
ISBN 0-7923-90932-6

Printed in the United States of America

To our parents,

Leung Lik-To and Choi Chik-Wan

Steven Leung

Norbert and Sally Shanblatt

Michael Shanblatt

Table of Contents

List of Figures

List of Tables

List of Code Segments

Preface

Beginning in the mid 1980's, VLSI technology had begun to advance in two directions. Pushing the limit of integration, ULSI (Ultra Large Scale Integration) represents the frontier of the semiconductor processing technology in the campaign to conquer the submicron realm. The application of ULSI, however, is at present largely confined in the area of memory designs, and as such, its impact on traditional, microprocessor-based system design is modest. If advancement in this direction is merely a natural extrapolation from the previous integration generations, then the rise of ASIC (Application-Specific Integrated Circuit) is an unequivocal signal that a directional change in the discipline of system design is in effect.

In contrast to ULSI, ASIC employs only well proven technology, and hence is usually at least one generation behind the most advanced processing technology. In spite of this apparent disadvantage, ASIC has become the mainstream of VLSI design and the technology base of numerous entrepreneurial opportunities ranging from PC clones to supercomputers. Unlike ULSI whose complexity can be hidden inside a memory chip or a standard component and thus can be accommodated by traditional system design methods, ASIC requires system designers to master a much larger body of knowledge spanning from processing technology and circuit techniques to architecture principles and algorithm characteristics. Integrating knowledge in these various areas has become the precondition for integrating devices and functions into an ASIC chip in a market-oriented environment.

But knowledge is of two kinds. While the ASIC revolution is made possible by the design automation endeavor, which captures the domain knowledge of physical design of a VLSI circuit in the form of CAD programs, high-level designs that involve applying knowledge in multiple domains consistently prove to be hard to automate. On the other hand, the application-specific nature of ASIC has intensified the need to synthesize information and knowledge in the three domains of algorithm, architecture, and technology.

This book is based on research that fulfills such a need by developing a design paradigm for the synthesis of knowledge in these areas. The underlying thesis of this research is that a decision-making perspective is conducive to a better understanding of the interactions among the three domains. From this perspective, a conceptual framework is presented to provide a logical view of the VLSI technology. Based on this framework, an ASIC architecture design methodology is developed, which is aimed at providing a focus for the decision-making process in each design phase. A paradigm, based on the architecture design of the IKS (Inverse Kinematic Solution) chip, is constructed to illustrate the design principles embodied in the methodology. The paradigm is, in a sense, an elaboration of what decisions are involved in a practical design and how these decisions are made.

A considerable amount of effort has been devoted to designing special computational hardware for robotics in recent years. As part of this effort, the paradigm is developed based on a well known algorithm to compute the inverse kinematic solution for a robot manipulator. The study of the characteristics of the algorithm and its interactions with various architectural concepts has led to a new computational architecture featuring a multiplier-accumulator with a cordic core (MACC). Evaluation has indicated that this new architecture achieves a higher performance and is more cost-effective than previous designs. Beyond its application in control, the potential use of ASIC in robotics is even greater in the latest undertaking of microrobot and sensor integration research. For researchers who are interested in these areas, the paradigm is instrumental in acquiring the essential knowledge for the effective utilization of this nascent technology.

While robotic control is chosen as the application area for the development of the paradigm, the primary target of this book is system designers and application engineers. During the course of this research, the DoD-sponsored VHSIC Hardware Description Language (VHDL) has been adopted by the IEEE as the Standard 1076-87. As the first industry-wide standard hardware description language, VHDL will have a significant impact on VLSI architecture design, an impact comparable to that of SPICE on circuit design. In view of this, the IKS chip design is described and simulated in VHDL. A concise introduction to VHDL as a programming language, as a design tool, and as a design environment, from a user's perspective, is included. With more than thirty VHDL program listings accompanied by detailed circuit diagrams, system designers will find this book both informative and useful for studying VHDL-based behavioral modeling techniques for IC designs.

Steven Leung

Acknowledgments

Research has always been an expensive undertaking, but in today's highly competitive world it is quickly becoming a luxury affordable only to a few lucky souls. In this respect, we are especially grateful for the financial support from the Dean's Distinguished Fellowship and the All-University Research Initiation Grant at Michigan State University, and in particular, the State of Michigan Research Excellence for Economic Development (REED) Fund. Without this support, this research effort, which has spanned over a period of more than five years since its initial conception, would not have been possible.

We are indebted to many people who have helped and guided us technically and emotionally in the course of this work. Among them, Professor P. David Fisher significantly contributed to several sections of this book, especially in the refinement of the conceptual framework for ASIC design presented in Chapter 3. Special thanks are due to Dr. Moon Jung Chung of the Computer Science Department at MSU for providing very helpful tutorial materials on VHDL. And finally, we are also grateful to Stephanie Shanblatt who has read through the manuscript and made many suggestions that have enhanced the book's readability.

Steven Leung
Michael Shanblatt

ASIC SYSTEM DESIGN WITH VHDL:
A PARADIGM

Chapter 1

Introduction

ASIC (Application-Specific Integrated Circuit) technology is revolutionizing the design, manufacturing, and marketing practice in electronics-related industries. The essence of this revolution is that in the pre-ASIC era, system designers dealt with microprocessors and standard integrated circuits (SICs), a design process of building systems from chips, whereas ASIC design is inherently a process of integrating systems into chips. This development necessitates a major change in the discipline of system design.

Central to meeting this ASIC challenge is a better understanding of the application requirements, algorithm characteristics, major architecture concepts, implementing technology, and, most important, their interactions [LeFS88]. While models, paradigms, or methodologies exist for each of these individual domains, their interactions lack any model or theory. The application-specific nature of the technology further complicates the effort to formalize these interactions.

The underlying theme of this research is that a decision-making perspective is conducive to a better understanding of the interactions among algorithm, architecture, and technology. That is, these interactions are best understood in terms of what design decisions are involved and how they are made in a practical design. Accordingly, the major thrust of this research is to develop an architecture design paradigm to expose these interactions. Principles behind the design decision making process can then be investigated. Understandings gained from the construction of the paradigm will facilitate future efforts in implementing application algorithms in ASIC hardware.

One application area that is of particular interest to this research effort is robotics. In recent years, advancement in technology has led to many novel applications. In these applications, sensory information is used extensively, and intelligence as well as real-time response is often required to cope with the unstructured environment or unpredictable event times. The desired enhanced

1

functionality for the next generation of robots — sensory information based adaptability and intelligence, locomotive ability, and real-time response — are creating new computation demands.

Recognizing the computational needs of robotics control, many researchers in the robotics community have been devoting considerable effort to developing special architectures for robotics applications [NiLe85, LeCh87, OrTs86, WaBu87]. These efforts, however, are based on the traditional SIC design approach. In the future, robot designs will become highly specialized and performance-oriented as required by novel applications. Robotic systems designed for these applications may not have an immediate large user base or a prolonged product lifetime due to their experimental nature. Such product requirements and market characteristics are matched best by ASIC technology. While the potential advantages of ASIC for robotics hardware design have been generally recognized [HeHK87], progress in adopting the ASIC technology has been slow. The development of an ASIC architecture design paradigm based on robotic control algorithms will help to speed up the transfer of this technology to the robotic community.

Because of its basic role in a robot control hierarchy, the algorithm for computing the inverse kinematic solution (IKS) is chosen for ASIC implementation. Among various approaches to computing the IKS, the closed form solution has the appeal of being well understood, widely applicable, typical of control algorithms in computation characteristics, and having known non-ASIC implementations. For these reasons, it was selected for special architecture implementation. By comparing the resultant design with previous implementations, advantages and tradeoffs of the ASIC approach can be better understood.

During the course of this research, the VHDL (VHSIC Hardware Description Language) has been approved by the IEEE as the standard HDL. This allows an architecture design to be documented independent of design approach and, to a lesser degree, technology. Moreover, documentation of a design in VHDL is formal since the design is described by a programming language (for which the syntax is formal) and its semantics can be verified by running the program. These features are highly desirable especially for a design being considered as a paradigm. Consequently, the IKS chip is described in VHDL and verified by extensive VHDL simulations.

1.1 Problem Statement

Future intelligent robots are expected to have the capability of making real-time decisions in an unstructured environment. The need to process a huge amount of kinematic and sensory information in real time under various environmental constraints renders special computer architectures necessary [LeSh88a]. Many robotic control algorithms, however, are characterized by large-grain computation with strong serial-dependence. Such characteristics favor algorithm-specific computational hardware implementation [LeSh88b].

The trend of algorithm-specific design is further reinforced by the advancement of ASIC technology, which makes it cost-effective to implement application algorithms directly in silicon. System design with ASIC technology, however, is fundamentally different from the traditional off-the-shelf SIC approach. A better understanding of the interactions among the application algorithm, various architecture styles, and the implementing technology is needed. In response to these needs, the aim of this research is to develop a paradigm based on the architecture design of an ASIC chip to compute the inverse kinematic solution for a robotic manipulator. By illustrating what design decisions are involved and how they are made, the paradigm will provide guiding principles for robotic engineers in mapping control algorithms in ASIC.

1.2 Approach

The general approach of this research is to 1) review design approaches and techniques of existing robotic computer architectures; 2) propose a model representing the essential aspects of ASIC design activities; 3) investigate the effectiveness of the model in characterizing computational characteristics of robotics algorithms through the experiments of implementing the IKS algorithm in ASIC; and 4) construct a design paradigm based on the experience of the IKS chip design.

The first task, an in-depth analysis and comparison of existing robotic computer architectures, serves three purposes. First, the advanced and useful design techniques can be adopted as part of the design repertoire. Second, critiques on previous designs enable the avoidance of known pitfalls in design decisions. Third, it is foreseeable that ASIC technology will coexist and compete with advanced microprocessor technology in the future. Thus, the study can help to identify areas in which the application of ASIC technology has the greatest potential for success. For the purpose of ASIC implementation, the review is focused on two key issues: how specific the architecture design should be, and what architecture style should be chosen. An understanding gained from the study will help to differentiate the more promising alternatives from those of less potential and thus set the stage for the IKS chip experiments.

For the second task, a model representing essential aspects of ASIC design activities is developed in accordance with the idea of a "frame" as a representation of knowledge. This leads to the notion of a conceptual framework, which consists of three knowledge frames — *Process, Hyperspace,* and *Repertoire.* The conceptual framework provides a logical view of the technology and thus allows engineers to acquire and accumulate VLSI knowledge systematically. In addition, the development of the conceptual framework represents a two-way interaction with the construction of the IKS chip paradigm. On the one hand, the IKS chip design is guided by understandings embodied in the conceptual framework; on the other hand, insights gained from the IKS chip design provide important feedback for improving the framework.

The third task involves designing an ASIC chip architecture for executing the IKS algorithm. The architecture design proceeds within the ASIC conceptual framework. Specifically, the IKS algorithm is characterized in terms of its position in the algorithm space. Architectural alternatives in functional units, interconnect topology, and control structure are identified and evaluated. Techniques classified in the design repertoire are applied with probable modifications. Analysis of the characteristics of the IKS algorithm leads to the architectural idea of a multiplier-accumulator pipeline overlapped with a CORDIC (COordinate Rotations DIgital Computer) core. The IKS algorithm is translated into a program written in pseudocodes assuming the MACC (Multiplier-Accumulator with a Cordic Core) datapath. The pseudocode program is then translated into a dataflow table and further manipulated using spreadsheet software. System timing diagrams, structural decomposition schemes, and communication links are concurrently developed and refined with the aid of the dataflow table. The dataflow table enables the designer to manipulate the computation flow of the application algorithm at the register-transfer level while maintaining a global perspective of the architecture. Once the interconnection scheme is satisfactory, the control signal patterns are specified and analyzed. An encoding scheme and the implementation of the control mechanism are determined. The control signal patterns are subsequently expressed as microcodes and further recast as the MACC instruction set. The microprogram, a microcode representation of the IKS computation, is then compiled from the dataflow table. The area and speed parameters of the resultant architecture are estimated based on data from the cell/functional module libraries of commercial ASIC products.

The IKS chip design is described in VHDL according to the circuit's structural hierarchy. Low-level building block circuits such as dynamic latches, multiplexers, shifters, counters, etc., are modeled and tested. Delayed time data are extracted from commercial gate array libraries and incorporated into the models. These building block circuits are then connected to form the next-level modules. Each of these modules is individually tested with regard to its functional specifications. The functional units and the storage elements are assembled to form the datapath, while various control modules are connected to form the entire control section. The timing aspects and the decoding mechanisms of the control section are verified. The datapath is then tested for the cordic operation and the instruction set. Finally, the datapath is combined with the control section to form the IKS chip. A C program is developed to generate realistic IKS input sets and expected outcomes. The VHDL description of the IKS chip is executed to simulate the IKS computation from input to output on different data sets.

The fourth task is the construction of the design paradigm based on the IKS chip architecture design. Central to the effectiveness of the paradigm is the illustration of interactions among the algorithm characteristics, the architecture styles, and the underlying technology in terms of architecture design decisions.

Therefore, the investigation of the actual design process and the resultant architecture focuses upon the following issues:

- What are the major architecture design decisions and their relative importance?
- What are the alternatives associated with each of these decisions?
- What are the tradeoffs of these alternatives?
- What are the determinant factors in making each of these decisions?
- What are the relationships of these factors and the algorithmic characteristics?

An understanding of these issues is presented in the form of an ASIC architecture design methodology. The IKS chip architecture design then becomes an example of the execution of this methodology.

1.3 Organization of this Book

In Chapter 2, an overview of the ASIC technology is presented first. This is followed by a discussion of issues in designing computer architectures for robotics. The discussion serves three purposes: 1) to present the field of computational architecture design for robotics in general; 2) to analyze how ASIC can be used in robotics; and 3) to review major advanced architectural concepts from the perspective of understanding their applicability in robotics. With this background, the basics of robotic kinematics and previous efforts on architecture design for computing the IKS are reviewed.

A conceptual framework for ASIC design is then presented in Chapter 3. The nature of design activities is first examined from a transformation perspective and a decision-making perspective. The conceptual framework organizes the broad range of IC system design knowledge into three categories: design process, design hyperspace and design techniques. Key concepts presented in the process frame include the hierarchy approach, the role of methodology, and a model representing the implementation of methodologies. The hyperspace frame articulates the role of the design space concept and outlines the framing of the architecture space and algorithm space as a means to facilitate recognition of design alternatives. The repertoire catalogs techniques for evaluating design alternatives. These three frames deal with different aspects of ASIC design, but they are integrated through an underlying theme of viewing design as a decision making process. That is, system designers must structure the design process so that the solution space is manageable and design alternatives are consciously sought and evaluated. Because of the growing importance of high-level design decisions, the discussion of these concepts will focus on one particular step — the transformation from task algorithm to architecture specifications.

Chapter 4 presents the architecture design paradigm of the IKS chip. It starts with a discussion of the assumptions and constraints of the IKS chip design effort. The design philosophy of this work is examined. The observation that architecture design in the ASIC environment is fundamentally different from that in the general purpose processor environment leads to the evolution of a design philosophy called DISC (Derived Instruction Set Computer). The fundamental principle of DISC states that the instruction set of an ASIC processor chip should be derived directly from its intended applications.

This design principle is embodied in an ASIC architecture design methodology developed from a decision-making perspective based on the notion of design space as delineated in the conceptual framework. The salient feature of this methodology is the division of the algorithm-to-architecture mapping into three phases with each phase having its own decision focus. According to the nature of each design phase, the decision focus is further defined into a form conducive to manipulation. Specifically, the functional unit profile, the dataflow table, and the control signal pattern profile provide the decision foci for the design phases of functional unit configuration, communication configuration, and control configuration, respectively. Decision alternatives in each phase are discussed.

The execution of this methodology is then illustrated by the architecture design of the IKS chip. The design is described in accordance with the three design phases prescribed by the methodology. The entire architecture design is concluded by an evaluation of the effectiveness of the IKS chip architecture in terms of architecture-level performance, area required, and resource utilization. Testability issues are also addressed.

Chapter 5 presents the VHDL description of the detailed design of the IKS chip. A brief introduction to the language is given first. The modeling objective and approach are explained. Issues involved in encoding a circuit design into programming semantics are addressed. Specifically, data typing as an abstraction mechanism is discussed, and a general delay time model based on timing description primitives provided by VHDL is presented. Certain component values of the delay time model depend on the overall design and the specifications of these values have to be delayed during the construction of the models. To facilitate the design process and simulation efforts, a programming strategy is developed, which employs the VHDL construct *configuration* to provide a well structured mechanism for specification of the delay time values.

The circuit hierarchy of the IKS chip has five levels. Following the top-down design approach, the functional tables of the chip's high-level circuit modules are derived from the dataflow table. The design of each circuit module is carried out to sufficient detail such that area and delay time parameters can be estimated from commercial gate array libraries. The simulation process, however, proceeds in a bottom-up fashion. In order for the presentation to provide some insights into both the design and modeling processes, a library is created for each level below the root, and circuit modules at each level and their related configurations are analyzed (a term referring to the process of translating VHDL sources into representations of an intermediate format) into the same library. The presentation of the design starts from the library Macros at the lowest level, which contains all the building block circuits callable from gate array libraries. Next, the circuit modules in the Level_2 library, which is in the middle of the circuit hierarchy, are described so that the presentation can follow the flow of the circuit decomposition. This also serves to illustrate how VHDL can be used to facilitate the top-down design process. The presentation then follows the bottom-up simulation path again until the root is reached.

All building block circuits and functional circuit modules are individually simulated to verify that the program semantics are consistent with the functional specification of each module. Starting at the bottom of the circuit hierarchy, tested modules are instantiated as components in a module at the next higher level until the MACC datapath is formed. The VHDL description of the datapath is then simulated to verify the cordic operation and the operation of the instruction set. Finally, the entire chip is simulated on realistic data sets from input to output. The results of these three simulations are presented.

In Chapter 6, a summary of the major results and contributions of this research is presented. Implications for robotics architecture design, ASIC system design, and CAD tool designs are cited. Future research issues are identified and discussed.

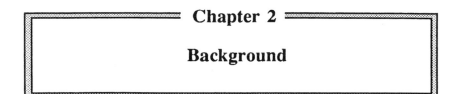

Chapter 2

Background

In this chapter, a brief overview of the ASIC technology is presented first. This is followed by a detailed discussion of issues in designing computer architectures for robotics. Previous efforts are reviewed from the perspective of understanding how ASIC can be applied to robotics. Specifically, two issues are addressed: how specific should an architecture design be, and what architecture style should be chosen? Then, the basics of robotic kinematics and previous efforts on architecture design for computing the IKS are reviewed.

2.1 The ASIC Challenge

The strategic significance of ASIC technology is often compared to the previous microprocessor revolution. To obtain some insight on this challenge, the basics of the ASIC technology are reviewed and the impact of ASIC on system design is examined.

2.1.1 ASIC Design Styles

The term ASIC conveys in its meaning a mixture of aspects of design approach, implementation technology, market orientation, and the subsequent product requirements. Currently, it covers semicustom designs including programmable logic devices (PLD), gate arrays (GA), standard cells (SC), and full custom (FC) designs. The dominant interpretation refers to GA and SC. Figure 2-1 illustrates the relative merits of these implementation styles. The major distinction between styles is the degree of design freedom in layout: FC and SC in masks (with the latter having some restrictions in cell height and locations for connections), GA in metal interconnects, and PLD in fuses. The greater the degree of design freedom, the greater is the design effort and the longer the design turnaround time. On the other hand, under the same fabrication technology, the achievable performance, measured by the functional throughput rate (FTR) in Gate·Hz/cm^2, decreases from FC to SC to GA to PLD. In practice, the performance difference between SC and GA is relatively unimportant compared with that between semicustom designs and SIC designs;

9

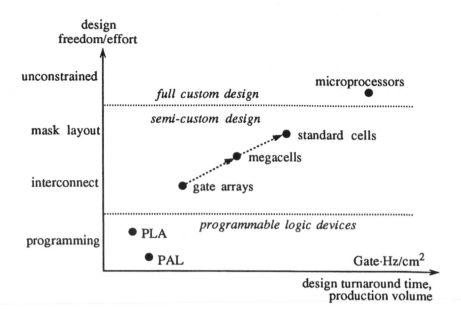

Figure 2-1. Relative merits of various ASIC implementation styles.

the determining factors in selecting a particular style are the acceptable design turnaround time and the projected production volume. For these reasons, GA has been the dominant choice for the past few years [Ber85]. In 1987, the industry was capable of fabricating arrays with a density of 35k usable gates using a 1.5-μm double metal CMOS process with over 200 I/O pins. Gate delays of less than 1 ns are typical [Lin87]. Furthermore, current sub-1.2-μm CMOS technology has made it possible to build application-specific chips with capacities of up to 200k gates [McL89]. To put this in perspective of computation power, note that a 16-by-16-bit signed integer array multiplier can be constructed with less than 2k gates.

Since most ASIC designs are implemented in either GA or SC, attention will be focused on these two design styles. In gate arrays, all levels of masks except the metal interconnections are predefined so that the wafer to a large extent can be prefabricated. This prefabrication of wafers is the main reason for the fast turnaround of prototype gate arrays. For standard cells, all masks are customized, but the height of each cell is fixed to reduce the design complexity. With the more advanced module generation design approach, both the height and width of the cells can be varied. With its limited design freedom, the chip size of the gate array is typically two to three times to that of handcrafted designs [Hur85, OkSG86, EBCH86]. Compared with standard cells using the same processing technology, the FTR of the typical gate array is smaller. However, due to reduced design and processing complexity, the cost of gate arrays is lower for small production volumes.

As indicated by the dashed lines in Figure 2-1, a trend appears to link the various approaches. LSI Logic, for example, offers a "Structural Arrays" technique that combines both gate arrays and megacells [Wa*et*85, WaRC87]. Other companies, like VLSI Technology, are offering gate arrays containing large standard cell blocks [McL86]. NEC is promoting a fully compatible gate array and standard cells package, and the computerized conversion of a design from gate arrays to standard cells. In contrast to the conversion approach, IBM has developed a design system that allows the complete intermix of standard cells and gate array functions. It can turn the unused cell locations into gate arrays in the background for personalization or for accommodation of minor design changes. Thus, by prefabricating masks, which mainly consist of large standard cell macros (front-end-of-line masks), the standard cell product can be manufactured in gate array turnaround time [Ho*et*87]. Some other companies, like Silicon Design Labs, offer silicon compilers using a module generation approach to generate megacells [BuMa85]. As a result, the boundary between various semicustom design styles has become blurred. It is anticipated that a complete path of gate array, to standard cells, and to unconstrained design will emerge in the next few years. This will provide a more satisfactory solution to a wide variety of demands in terms of cost, performance, turnaround time, and production volume. An immediate implication is that design decisions regarding tradeoffs of implementation styles will become relatively unimportant since migration from one style to another can be achieved with much less effort. This trend underlines the importance of an integrated design environment.

2.1.2 The Cutting Edge

Forces that motivate the technological advancement of ASIC include:

- Better performance;
- Higher reliability;
- Lower non-recurring cost;
- Faster design turnaround time;
- Tighter design security.

The first two of these driving forces are also applicable to general IC technology; the last three factors contribute to the popularity of ASIC. These factors are further examined in the remainder of this section.

Better Performance

The demand for better performance has always been a major concern of the semiconductor industry. Currently, however, performance gain from processing technology is diminishing [NeVi87]. Clock speeds, the figure of merit in performance measure, of silicon microprocessor-based systems are limited to 50 MHz for off-the-shelf components. Moreover, in the SIC design environment, performance is further limited by compatibility concerns or the need to support the existing technology base. The addition into hardware of system-supporting functions, such as memory management and exception handling functions, imposes a penalty on performance in the form of system overhead per instruction [Hen84]. In contrast, such problems have less bearing on ASIC designs. In fact, the fundamental performance advantage of ASIC design is often not due to the more advanced processing technology (resulting in, for example, faster gate delay times), nor is it due to some exotic computation architecture. The advantage of ASIC is mainly derived from the fact that because of the application-specific orientation, many system overhead functions usually associated with conventional microprocessors, or even RISC (Reduced Instruction Set Computer) processors, can be totally eliminated. Hence, with simpler control as the norm, the possibility of achieving very high performance through, for example, pipelining is greatly enhanced. Of course, this higher performance is measured with respect to the targeted applications, and is achieved at the expense of flexibility offered by general purpose micro-processors. This fundamental attribute of ASIC design emphasizes the importance of high-level design, particularly at the architecture level, for a given task algorithm. Additional advantages include the reduction of the overall system hardware size, which may be critical in such applications as improving the locomotive ability of a robot or in airborne systems.

Higher Reliability

Reliability is a complex matter. On the positive side, reliability at the system or board level decreases exponentially as the number of components increases. Because of the higher logic capacities, semicustom chips can replace a moderate to large number of standard chips and thus improve the system reliability. On the negative side, when existing designs are implemented in ASIC, previously accessible nodes for testing may become inaccessible. To maintain the same quality level, measured by the percentage of undetected defective products, requires a higher quality chip [Mey86]. This requirement translates into more stringent testing requirements of ASIC designs in general. But overall, current advancement in IC processing, packaging, and testing technologies have made greater improvements in IC reliability. For example, the failure rates for molded linear ICs under the industrial standard 168 hour burn-in test has been reduced from 21.48% in 1979 to 0.22% in 1982 [Pan86]. This is a two order-of-magnitude improvement in less than five years! As advanced fault-tolerant and testing techniques are incorporated into IC design, the reliability can be expected to increase. Therefore, by reducing the number of components at the board level through integration and by profiting from advancement in processing and testing technology, the ASIC approach offers a double gain in overall system reliability.

Lower Non-recurring Cost

According to a 1986 survey, non-recurring development expenses for a 3,000-gate digital IC ranged from well under $10,000 to over $100,000 [VSDS86].

Further reduction in the cost of prototyping is expected. This will open up new opportunities for rapid prototyping and for the development of products that have either a short lifetime or require only a small production volume. But the cost factor of IC design must be examined in the context of the entire product cost at the system level, which includes cost of PCB design, assembling, and field maintenance. The benefit of ASIC in reducing the system cost, which accounts for 80% of the total product cost, is dramatic (due to smaller size, fewer components, less power consumption, reduced assembly cost, etc.) and often outweighs the development cost [FePa87]. A recent study has identified the level of integration (LOI) measured by gates per pin as the key determinant of the total IC-related cost, and ASICs can raise the number of gates per pin from 2 in MSI/SSI to a range of 40-200. The study further establishes that gate arrays have a lower cost than the MSI designs with build volumes as low as 1,000 devices if they replace at least 5-10 equivalent MSI/SSI circuits [FePa86, FePa87]. In fact, replacement of 20-60 equivalent ICs with current gate array technology is not unusual. A most striking conclusion is that when all cost variables are taken into account, significant cost-reduction opportunities lie in the virtual elimination of all MSI in system design.

Faster Design Turnaround Time

From an economic standpoint, the timing of new product introductions into the market is often critical. Such a timing requirement is beginning to be satisfied with the semi-custom design approach. Faster turnaround time for ASIC production can be attributed to three reasons. First, most present day ASIC applications are implemented with either gate arrays or standard cells. For gate arrays especially, the processing time is significantly reduced since only metal interconnections need to be processed. Silicon foundries now advertise 15-day guaranteed service for CMOS 2-layer metal fabrication. The second reason is that sophisticated software design tools and CAE systems have been developed, which reduce the length of the physical design phase by reusing large portions of proven designs. For example, LSI Logic offers a compiler that can incorporate a 70-ns, 32-bit multiplier building block into the user's system and complete all design tasks, from routing, timing simulation to layout generation, within 48 hours [Ber85]. Finally, the third reason is that less stringent requirements, resulting from a reduced burden of system supporting functions, are also helping to shorten the design cycle.

Tighter Design Security

Implementing task algorithms in ASIC hardware offers an effective leverage in controlling design security. Even though "reverse engineering" is always possible (particularly in the case of gate arrays) it still poses a considerable obstacle to piracy. This is especially true when compared with duplicating the software/firmware of, for example, microprocessor-based designs. By custom-designing critical system components, developers now can control the security of the architecture and algorithm. This factor may play an even more important role in the future.

2.1.3 Impact of ASIC on System Design

The impact of ASIC on system design can be better understood by examining the interaction patterns among the three domains of algorithm, architecture, and technology from a historical perspective. Figure 2-2 illustrates the evolution of these interactions from the pre-VLSI era to the ASIC era.

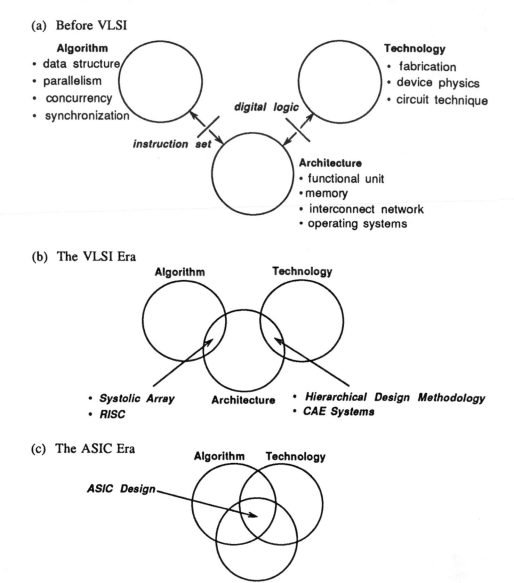

(a) Before VLSI

Algorithm
• data structure
• parallelism
• concurrency
• synchronization

digital logic

instruction set

Technology
• fabrication
• device physics
• circuit technique

Architecture
• functional unit
• memory
• interconnect network
• operating systems

(b) The VLSI Era

Algorithm   **Technology**

• **Systolic Array**   **Architecture**   • **Hierarchical Design Methodology**
• **RISC**   • **CAE Systems**

(c) The ASIC Era

Algorithm   Technology

ASIC Design

Architecture

Figure 2-2. Interactions among algorithm, architecture, and technology.

In the pre-VLSI era, the boundaries of these three domains were quite clear-cut. On the one hand, digital logic served well as an interface between architects who build systems from SICs and the process engineers who worked out the implementations of the circuits and devices. On the other hand, the instruction set also served well as an interface between architects and computer users who developed algorithms to be executed on machines whose behaviors are represented by the instruction sets.

As IC technology progressed to the VLSI era, the feasibility of integrating a large number of devices on a single chip established closer interactions between these three domains. On the architecture-technology side, the evolution of hierarchical design methodology as a means to control the complexity and the emergence of powerful CAE systems occurred. On the algorithm-architecture side, first systolic arrays and later RISC appeared as innovative architecture concepts that attempt to take advantage of both the architecture and certain algorithmic characteristics.

The advent of the ASIC era may be visualized as the formation of the overlapping region as illustrated in Figure 2-2(c). Compared with the microprocessor revolution in the pre-ASIC era, it is striking that system designers now must master a body of knowledge much larger than what was adequate previously. Indeed, integration of circuits is above all an integration of knowledge from the fundamentals of material properties, processing technology, device characteristics, circuit techniques, high-level abstractions of logic, and structural organization (architecture) principles [Seq83]. Integrating knowledge of this magnitude presents a tremendous hurdle to system designers and has created an increasingly wider gap between VLSI designers and application engineers. This gap is most plainly manifested as the disparity between our capability to fabricate and our capability to design [May85].

Moreover, system-level designs now involve users, CAE tools developers, and ASIC vendors. Interactions between these three parties are so complex that currently only the interface between CAE tools and silicon foundries, through the ASIC vendors, has begun to be standardized. The evolving nature of ASIC technology and the diversity in application areas, design tools, design approaches, and fabrication processes make it difficult to define clean interfaces between the three parties. Obviously, unlike the case of microprocessors, broadening the use of ASICs to new applications requires extraordinary effort. This effort, as made clear by the examination of the evolutionary development of the IC technology, must focus on attaining a better understanding of the interactions between application algorithms, architecture, and technology.

2.2 Computer Architecture Design for Robotic Control

In the past few years a wealth of knowledge about computer architecture design for robotics has accumulated. Among the proposed designs, however, only a few have been implemented in labs and even fewer have appeared in commercial products. While previous computational needs are being satisfied due to advances in both design and technology, new applications continue to create new demands. The present condition of the robot market has forced researchers to rethink the future direction of the field and, as a result, a new research agenda has been proposed [Whi86]. In addition, nascent developments in ASIC technology have created new opportunities and challenges in architecture design. It now seems appropriate to reexamine previous efforts in robotic computer architecture designs.

Underlying the forces of market trends, ASIC technology, and the new robotics research agenda is the notion of "application specific". In this spirit, some trends in robotics applications are examined and issues involved in hardware implementations are considered. For the purpose of ASIC implementation, this study is focused on hardware designs for control. Selected previous works are then classified with respect to architecture types, design features, and performance. Based on the observations on design and application trends, design techniques, and architecture performance, the issue of how "specific" the architecture should be is addressed. Background on computer architecture design and the characterization of interactions between algorithms and architectures is then presented. With this background, four major architecture concepts — pipelining, RISC, systolic arrays, and multiprocessors — are reassessed for their applicability in robotics.

2.2.1 Trends in Architecture Design for Robotics

The use of robots in industry has been focused on improving productivity through the automation of manufacturing. In the early 1980's, progress in computer technology and AI (Artificial Intelligence) research generated high expectations of furthering this goal by introducing "intelligence" into robots. However, the 1982 forecast of an imminent robot boom has failed to become reality. In 1986, the sale of robots in the US reached only $441M (6,219 units), far less than what was projected [Roo87]. It has become clear that in industrial applications, the use of robots is only one of the factors in the automation endeavor. Indeed, concepts such as design for assembly, environment structuring, and design-manufacturing integration might have much more to do with productivity than making robots intelligent [Whi86, Whi87].

At the same time, advancement in the technology has led to many novel applications. Robots are now operated in clean rooms, hospitals, nuclear plants, the deep seas, and in the future, in space stations. They are being used to inspect IC circuits, assist in surgical operations, perform surveillance and maintenance jobs. They are expected to supervise the fabrication of supercomputer chips in space and to help the elderly and disabled back on earth. Sensory information is used extensively in these new applications. To cope with the unstructured environment or unpredictable event times, intelligence and real-time response are often required. In contrast with traditional industrial applications, many of these new applications are relatively cost-insensitive but require robots with high performance and special features.

The differentiation of robotic R&D into these two directions has a profound impact on robotic computer architecture designs. In next section, some of the technical and economic implications are explored.

2.2.2 Dedicated Hardware Implementation Considerations

Recognizing the computation demands for next generation robots, considerable research effort has been devoted to the development of special architectures for robotics. Future computation requirements, however, must be understood not only in terms of individual application needs, but also in the context of the overall system organization. Accordingly, a three-layer hierarchical computation model has been developed for a generic robotic system to provide a global view, permitting assessment of the computational needs for control (manipulative and navigational) and sensory information processing. The economic implications of the application trends, market characteristics, and computational requirements on special hardware implementation are discussed.

2.2.2.1 The Robotic Computation Hierarchy

Functions performed by the computer in modern robotic systems are hierarchically organized. Different computational needs for different applications (e.g., manipulative vs. navigational) must be examined in the context of the hierarchy in order to obtain an economical architectural solution. How the hierarchy is organized strongly influences the way it is implemented. For a technology as evolutionary as robotics, important considerations include allowing variation without jeopadizing standardization, and encouraging growth without destroying stability. These conflicting requirements are not unusual, particularly for a technology still in its infancy. However, there is a successful precedent in handling such a situation: the computer networking technology. The key of that success seems to lie in the fact that the computer networking hierarchy is organized by layers, which allows greater variations, in an open-architecture fashion. With this in mind, a robotic computation hierarchy model organized in a similar fashion, as illustrated in Figure 2-3, is presented.

In this model, the robotic functions are partitioned and organized into three layers: *management*, *reasoning* and *device interaction*. The top layer of management is mainly responsible for various routine tasks such as resource allocation, coordination, and user interfacing. The bottom layer of device interaction is directly coupled to various devices, which can be divided into the two domains of manipulators/joint-actuators and sensors. The functions performed by the computer corresponding to these two domains are control and sensory information processing, each having its own hierarchy. A middle linking layer of reasoning consists of tasks such as world modeling, temporal and/or geometric reasoning, causal analysis, decision-making, and planning.

With this model, a specific robotic computation system can be viewed as an instantiation driven by a particular application. In such an instantiation, certain components of the model robot can be absent or assume different importance. On the other hand, new application layers can be added on top of the

User		

Layer 2
 	
Management
	• user interface
	    — programming support
	    — graphic/iconic protocols
	• operating system environments
	• resource allocation
	• coordination	

Layer 1 *Reasoning*	• decision making • causal reasoning • temporal reasoning • geometric reasoning • planning • world, modeling	
Layer 0 *Device* *Interaction*	○ control • adaptive control • force compliance • dynamics • kinematics	○ sensing • multisensor fusion • interpretation • feature extraction • preprocessing
	manipulators	sensors

Figure 2-3. The three-layer computation hierarchy of a generic robotic system.

management layer to provide sophisticated functions for developments such as simulations, FMS (Flexible Manufacturing System) or CIM (Computer Integrated Manufacturing).

2.2.2.2 The Computational Needs

In the three-layer robotic computation hierarchy, tasks in Layer 2 belong to the domain of operating systems and thus are of a more general nature. Tasks in Layer 1 are largely in the research stage and not well understood yet. Since the purpose of this research is aimed at hardware implementation, the investigation of the computation needs is focused on control and sensory information processing.

The computational needs in control can be further divided into manipulator control and navigation control. In manipulator control, a sampling frequency of 60-100 Hz is considered adequate due to the mechanical constants of the arm; however, a sampling frequency as high as 5 kHz is anticipated [WaBu87]. The number of arithmetic operations (multiplication and addition) required in each cycle for a six-DOF (Degrees-of-Freedom) manipulator varies from under 1,500 with the inverse kinematics (Newton-Euler formulation) to over 6,000 with the resolved motion adaptive control method [LuWP80, LeLe84]. If performed with floating-point operations, the latter figure corresponds to a throughput requirement of 30 MFLOPS for a latency (response time) of 200 µs. Required types of elementary functions, such as trigonometric function and square root, vary from algorithm to algorithm.

The computational needs for navigation are less understood, but experiments confirm that enormous computation power is needed for perception processing and image-based world modeling. Also, the necessity for real-time response is obvious. Current experimental autonomous navigation vehicles can move at only a few kilometers per hour, and yet the on-board computer is already overwhelmed [Ada86, GoSt87]. If the navigation speed is to increase by an order of magnitude as planned, then the computation power must be increased even more due to the more stringent requirements for many concurrent control processes. Moreover, additional constraints are placed on the size, weight, and power consumption of the computation hardware because of the effect of payload on mobility.

Sensors that require special computation processing are associated with two particular types — visual and tactile. For visual sensing, if one image operation requires a simple operation on each pixel in an array of 256×256 gray level pixels, then a computation throughput of 2-4 MIPS is required at 30-60 Hz video frequency. Higher resolution with color information, combined with more complex image operations, can easily drive the throughput requirement into the billion operations per second level [Mon86]. Tactile sensor research is presently in a primitive stage with major emphasis in searching for durable and robust materials. While the need for locally processing the sensor signals is generally recognized, the mechanism to do this is a subject of controversy [Har82, RaTa82, Gol84].

2.2.2.3 Economic Considerations

The economics of designing special computer hardware involves two main factors — market and cost. According to the RIA (Robotic Industries Association), the number of US industrial robots installed in 1986 reached 25k. The market sale is projected to reach $2,132M (125k units) in 1990 [Roo87]. As late as 1984, 80% of robot applications fell into just a few categories such as welding, spraying, machine loading, and material handling [ShGS84]. The auto industry has long been the primary user, although its market share is expected to decline from 60% in 1983 to 30% by 1989 [Coh85]. While the robot population is small, the number of species is large and growing. A recent robot database contains 220 industrial robots with payload from under 10 lbs to over 1,000 lbs and horizontal/vertical reach from a few inches to over 10 feet [McD85].

Another element in market consideration is the product lifetime. It is unrealistic to expect that a computational hardware design will last for decades since robotics and the related technology are still in a rapidly evolving state. In fact, the average product lifetime of the Sx controller, a popular controller for industrial robots from ASEA, is only about 6 years and shortening. The total number of S2 installations in its entire 6-year lifetime is 5,000 [Kee86]. In short, the robot market is characterized by small production volume, wide diversity, and accelerating dynamics. Predictably, as long as the so-called personal robot for household chores is not technically and economically feasible, these market characteristics are unlikely to change dramatically.

On the other side of the equation is the cost. For any specially designed hardware, the design cost rather than the production cost is usually the dominant component. For this reason, special IC designs for robotics have been out of the question in the past. But this situation is changing as ASIC technology becomes more prevalent. Non-recurring cost of prototyping ASICs currently runs about $20k for low-complexity designs and is expected to decline further [HiRa87]. In fact, the latest S3 controller contains several custom-designed LSI circuits. A recent study has established that gate array designs can be cost-effective even with production volumes of less than 1,000 units [FePa86]. Moreover, the cost of computer hardware is only part of the total cost of a robotic system. The benefits of using ASIC parts at the system level can outweigh the higher development cost. The time-to-market and performance features become the two most important factors in determining a product's success as the market moves in the application-specific and performance-oriented direction. The custom design approach, even at the IC component level, can be justified since these attributes can be met.

As can be seen from the previous discussion of application trends, new demands for computation power come primarily from novel applications. Robotic systems designed for these applications are highly specialized, experimental, and evolutionary in nature, and thus do not seem to have an immediate large user base. Hence, the strategy of designing a generic architecture encompassing the entire robotic computation hierarchy may not pay off, since the product requirements are so diverse and the total market is small and dynamic. Furthermore, when special hardware design is considered, at least one of the following conditions is usually present: either the tasks at hand are well understood (thus the algorithm that the hardware implements is relatively stable), or the required performance exceeds the capability of existing hardware. Taking all these factors into consideration, it appears premature to expand hardware design efforts beyond the device interaction layer. But even within this layer, how general (or specific) should an architecture design be? Before answering this question, an examination of previous efforts is useful.

2.2.3 A Survey of Previous Work

In light of ASIC hardware implementation, this study concentrates on architectures for control functions in the device interaction layer only. As a result, architectures for image processing and a number of systems designed for

Table 2-1. A Comparison of Computer Architectures for Robotic Control.

Author	Intended Application	Type	Design Level	Funtional Units	Architecture Communication	Characteristics Control	Performance
KIWa82*	Hexapod (6 legs) control	M	System	5 LSI-11/03 processors (board)	loosely coupled, fully connected	centralized	main cycle 117ms, $S = 3.47$, $Ea = 0.87$
Gol84*	Hand cont., Touch sensing	M, P	System, Processor	8,16-bit µ-Proc. MC68701 based sensor interface	4 hier. levels dedicated links	hardwired control store	target cycle time 10ms (3 fingers 2 joints & 3 sensors)
Nas85	Linear algebra	S	System, PE, Instr.	14 X16 PEs with custom PE chip	2-D mesh	lockstep synchronization	100–500 MIPS
NiLe85	Inv. dyn. (NE)	M	Conceptual	6 processing modules	1 global bus 1 local bus	centralized, microprogram	$S = 2.5$, $U = 0.57$, $Ea = 0.42$
OCOS85	Jacobian, inv. dyn.	M	System	N+1 processors custom FPM/FPA	1-D pipeline	lockstep	custom chip at 2 MHz, $S < 1.5$, $Ea < 0.2$
KaNa85*	Inv. dyn. (NE)	M	System	N 8086/8087, 32k RAM, 16k ROM	shared global memory	centralized	10 MFLOPS/PE (peak)
NSHB86	Hand control	M	System	9 68000s, 4.5 MB mem. 320 ADC, 40 DAC	Multibus, message passing	distributed	9 MIPS (nominal)
OrTs86	IKS	P	Processor	32-bit FPU	3 busses	microprogram	2,389 FPU cycles
LeCh87	IKS	M	System	25 cordics, each dedicated	pipeline mapped from algorithm	hardwired control	$U = 1$, 40 µs (TR), $S = 1.38$, $Ea < 0.06$
LeSh87	DKS	P	Reg./Logic	18-bit MAC	dual bus	hardwired	73 cycles (6 DOF)
JaOr87	Inertia matrix	S	System, PE	32-bit FPM+FPA, 4 16-bit ports	1-D, 1 global bus; or 2-D mesh	microprogram	2,389 FPU cycles, $Ea < 0.2/N$
BeSz87*	Telerobot control	M	System, Processor	NS32016+FPU, 128k RAM, 32k ROM	shared memory via Multibus	3 hier. levels, control store	control 6 DOF systems at 1 kHz sampling rate
WaBu87*	Control	M	System, Processor, PE	custom robot proc.: 3 TMS320 32-bit DSP 1 cordic chip, 64 vector registers	VME bus for host private bus for devices, shared memory	control prog. from host, 3-stage instr. pipeline	3 X3 MV multiplication in 900ns (TR) at 10 MHz, IKS+DKS in 0.25 ms

Note: P– Processor, M – Multiprocessor, S – Systolic array, TR – Throughput Rate, FPM(A) – Floating Point Multiplier (Adder);
* Known to have been implemented.

support of programming and research environments are not included. The main focus is on the design approach and architecture styles.

A total of 13 designs for robotic control functions are reviewed. They constitute a collection of architectures rich in intended applications, architecture types, and design approaches. The selected works are compiled in Table 2-1. Information on performance, functional units, and the communication topology is based strictly on published data, while that of the control scheme in a number of cases is based on this author's assessment.

Among the 13 designs, over half are multiprocessor systems, especially in the hand control area. Most single processor systems are designed for specific control algorithms. Two designs are implemented in systolic arrays. In the remainder of this section, design features are summarized, followed by an analysis of the question of how specific an architecture design should be.

2.2.3.1 Design Features

Among the 13 designs, most are based on off-the-shelf functional units, but a number of architectures involve designing special IC chips, sometimes as a component of the target system. And among those off-the-shelf SIC-based designs, the emphasis is shifting from the first generation general-purpose microprocessors to the second generation special functional units such as digital signal processors (DSPs), floating-point processors, or cordic processors.

The intended applications increasingly appear targeted to a specific algorithm. This trend is perhaps due primarily to the complexity of the algorithms involved, and secondarily to the generally decreasing hardware cost, which makes a narrower application target economically feasible. This development is in tune with the market trend of "application specific". In parallel with the narrower target application, more designs are for processors, in contrast with stand-alone systems, and the design level is correspondingly shifted down the design hierarchy. From the perspective of the robotic computation hierarchy of Figure 2-3, the strategy of designing processors for the device interaction layer and attaching them to general purpose computers that handle the upper two layers' computation makes sense in the near-term.

Except for systolic arrays for image processing, many of the multiprocessor designs are not scalable or expandable. In fact, most of these designs are targeted for implementations that incorporate no more than ten processor units. Also, development support is generally inadequate except in a few cases where commercial interests play an important role.

2.2.3.2 Design Approaches

In designing robotic computation hardware for the device interaction layer, the commitment of an architecture implementation can be classified into four levels of increasing specificity as follows:

- Domain-specific — specific to a particular domain (i.e., control or sensing) but general to all tasks within the domain;

- Task-specific — specific to a particular task within a domain (e.g., IDS, IKS, etc., within the control domain);

- Algorithm (approach)-specific — specific to a particular approach to a task (e.g., algebraic or iterative approach of the IKS);

- Robot-specific — specific to a particular robot or robot family (actuator type, joint type, etc.).

For an architecture design specific to a particular level, it is assumed that levels lower than the specified one will be covered. For example, a design for IKS specific to the task level means that it can handle the IKS computation of various approaches for any robot. When the commitment level moves from the robot level to the domain level, the hardware complexity generally increases due to the increasing demands for flexibility. In practice, if the architecture is dedicated to a robot or robot family, the robot's mechanical and geometrical characteristics will favor certain algorithms. Hence, the difference between the robot level and algorithm level is relatively minor. On the other hand, the shift from algorithm level to task level usually requires different types of operations (trigonometric, square root, etc.) and a more elaborate control structure. And if a design is already committed to the task level, the increase in hardware complexity due to the shift to the domain level will be modest. With this in mind, it is not surprising to see that most of the designs listed in Table 2-1 are specific to either the algorithm or domain level.

Several factors seem to favor the commitment to the algorithm level for control implementations. First, flexibility and performance are always antagonistic. However, performance is emphasized from the application-specific orientation and this requirement is easier to satisfy through, for example, hardwired logic, if the design is committed to the less flexible algorithmic level.

Second, the application requirements in positional accuracy, repeatability, payload, and real-time response tend to govern the choice of robots and subsequently affect the choice of a particular approach. For example, the robot used for inspecting an IC circuit will be very different from the one for material handling; consequently their requirements in accuracy, payload, and sampling rate will also differ. However, even though the requirements differ widely, for practical reasons they will not be in a continuum but rather will cluster into several groups evolved from a score or so of robot families. The actual requirements have a natural bias to the algorithm-specific approach since algorithm development is strongly influenced by the robot design.

Third, robotic control tasks are characterized by a considerable degree of decoupling between tasks but high serial dependency within a task [NiLe85]. Because of this decoupling effect, concurrent execution of these tasks is possible. Since the overall computation must be done in the sampling cycle, the single system, general-to-all-task approach will aggregate the performance requirements. An algorithm-specific design approach, on the other hand, allows relegation of the overall performance requirements into several components, each with relaxed requirements. Since each component is tailored to a specific algorithm, higher overall hardware utilization can be achieved. Also, this approach represents a truly modular design style, which will facilitate flexible configuration of systems for particular needs.

Fourth, when the architecture is designed for general robotic algorithms, the hardware is more complex and the demand for software development support grows. In contrast, with algorithm-specific architecture designs, the design effort is less, not only because of the simpler hardware, but also because the demand for software support is minimal or even unnecessary.

Finally, designs with lower hardware complexity will have a definite edge when ASIC implementation is considered. The small production volume characteristic of the robot market renders the design effort the major component of the development cost. In this context, the choice between the domain-specific approach and the algorithm-specific approach represents the choice of where to direct the design effort. In the former case, it is a mixture of software, firmware, and board-level hardware. In contrast, the algorithm-specific approach allows a larger part of the design effort be devoted to the chip-level hardware, and thus has a greater potential for exploiting the ASIC advantage. When all these factors are taken into account, the algorithm-specific approach is the most advantageous choice.

2.2.4 Matching Architecture Styles to Algorithm Characteristics

Hon and Reddy have pointed out that the efficient implementation of an algorithm on a particular machine is largely shaped by the architecture of that machine. On the other hand, an architecture type that favors a certain kind of computation may not be as efficient when other computations are required [HoRe77]. In the application-specific design environment, the question of exactly what type of architecture is suitable for what type of algorithm can be examined in the context of robotic applications. The metrics for measuring the effectiveness of a design with respect to the execution of application algorithms is presented first. Four major advanced computer architecture concepts are then reassessed.

2.2.4.1 Metrics for Evaluating Architectural Effectiveness

A computational system is composed of one or more data processing units organized for executing algorithms. For the purpose of this study, three types of systems — single processor, multiprocessor, and systolic array — are of particular interest. The term concurrent refers to simultaneous execution of processes belonging to different jobs, while in parallel systems, the processes being executed belong to the same job.

In either single or multiple processor systems, the architecture is characterized by functional components, communication topology, and control structure. The design of a computer architecture thus involves specification of these three aspects. They can be specified at various levels such as system (processor-memory-switch), processor, register-transfer, or logic. Specifying architectural aspects at a lower level entails greater design freedom, but requires more design effort. In this review, PEs (Processing Elements) are distinguished from processors. The former are usually used in systolic arrays and do not have complex control structures. Furthermore, a few designs address the data flow characteristics of an algorithm and the specifications of the architectural aspects are more or less of a conceptual nature. Thus, the design level of these architectures are classified as conceptual.

If a design is for a special purpose such as the execution of a specific algorithm, then the specifications should be derived from the algorithm so that the resultant architecture will best match the characteristics of the algorithm. Unfortunately, the mapping between algorithms and architectures is not one-to-one and architecture design still largely depends on experience and intuition. Hence, it is desirable to have some metrics for measuring how good an architecture matches a particular algorithm. Care, however, must be taken in interpreting these metrics. It is necessary to distinguish metrics for specifying computational requirements from those for measuring design quality. A design that scores high in one category does not necessarily infer the same good quality in the other. Only the design that satisfies the computational requirement with good design quality is a cost-effective one.

Computational requirements are usually specified in terms of throughput rate and latency. The throughput rate measured in MIPS (Million Instruction Per Second) is often used to indicate the raw computation power of a system. But the MIPS number is usually evaluated according to the "average" machine cycles per operation (or instruction) based on some benchmark programs. Furthermore, while the efficiency of the compiler (i.e., the optimization technique and the ability to take advantage of special features of a particular machine architecture) is important, it is rarely reflected in the MIPS numbers. Therefore, when a system is designed for the execution of a specific algorithm, the MIPS number is not useful. When taking into account the I/O problem, the number of solutions per second in matching the sampling rate of the target robotic system is a more accurate specification. Moreover, the throughput rate should not be used to extrapolate how fast the solution can be computed. This is specified by latency, the time elapsed from the input of the operands to the time when the result is available.

A popular metric for evaluating the effectiveness of a design is the resource utilization factor U, defined as the ratio between the total usage time over the total available time of the measured resource. The utilization factor depends on how the resource is defined and sometimes may be confusing. For example, the utilization factor for a MAC (Multiplier-ACcumulator) treated as a single resource can be higher than the same design with the multiplier and accumulator treated as separate resources. But in the former case, the clock cycle may be longer, which can result in a longer latency. Additionally, in pipeline designs, it is not uncommon that the utilization is high and yet little improvement is obtained in latency. Thus, a more indicative metric is the speedup factor, S. For single processor systems, the speedup can be obtained only by comparing the speed with other designs. For multiprocessor systems, the speedup is defined as the ratio between the execution time using a single processor and that of using multiprocessors. Amdahl has pointed out that the speedup is limited by the amount of parallelism inherent in the algorithm, which can be characterized by a parameter f, the fraction of the computation that must be done serially [Amd67]. He thus reasons that the maximum speedup of a P-processor system in executing an algorithm as a function of f is given by

$$S_{max} = \frac{P}{fP + 1 - f} .$$
(2.1)

Note that $S_{max} = 1$ (no speedup) when $f = 1$ (everything must be serial), and $S_{max} = P$ when $f = 0$ (everything in parallel).

Based on this insight, two different metrics can be developed for a given design. The absolute effectiveness E_a given by

$$E_a = \frac{S}{P}$$
(2.2)

is an overall cost-effectiveness measure for an architecture design. On the other hand, if the f factor of an algorithm is known, the relative design effectiveness denoted by E_r can be obtained from

$$E_r = \frac{S}{S_{max}} .$$
(2.3)

E_r can be thought of as an indicator of the design quality. Note that if we assume $E_r = 1$, then E_a is the maximum, which can be used to assess how suitable a particular architecture style is for the given algorithm. Essentially, E_a is limited ultimately by the inherent parallelism of the algorithm, while E_r can be viewed as a measure of design ingenuity.

While the absolute effectiveness E_a is treated as the major indicator of how suitable an architecture style is for a specific algorithm in this study, two main factors will limit the general applicability of the analysis. First, in eq. (2.1), f is assumed to be something inherent in the algorithm, but in fact, the actual value is somewhat implementation dependent, (for example, the number of processors used). Eqs. (2.1) and (2.2) can be rearranged to obtain f as a function of E_a and P as

$$f = \frac{E_a^{-1} - 1}{P - 1} .$$
(2.4)

If "cost-effectiveness" (E_a) is fixed, then f obtained as a function of P represents the cutoff serial fraction above which an algorithm-architecture pair cannot achieve the desired cost-effectiveness. It is easy to see that this cutoff value decreases rapidly as P increases. Intuitively, if the number of processors increases, then the algorithms executed on that architecture must have a greater degree of parallelism to maintain the same level of cost-effectiveness.

Second, most of the resource utilization in SIC designs refers to the processor or functional unit utilization only. Actually, interconnections occupy more silicon area than functional devices in VLSI and the bus bandwidth usually sets the ultimate limit on performance. Without taking the utilization of interconnections and busses into account, the implementation of the same architecture in ASIC may not be cost-effective. Understanding these limitations can help to reduce the possibility of misinterpreting the performance data.

2.2.4.2 Assessment of Advanced Architecture Concepts

In this section, four major computer architecture concepts — pipelining, RISC, systolic array, and multiprocessing — are reassessed for their applicability in robotics based on the performance data of the architecture designs reviewed in Section 2.2.3.

2.2.4.2.1 Pipelining

Pipelining refers to the partitioning of a process into successive, synchronized stages such that multiple processes, each in a stage different than others, can be executed in parallel. Depending on the granularity of the process, three types of pipelining techniques can be identified. For instruction pipelining (medium grain), the process of the instruction execution is often partitioned into stages of instruction fetch-decode, operand-fetch, and execution. Intra-functional unit pipelining (fine grain) divides the execution unit (usually a combinational circuit) into several segments of equal delay time. Inter-functional unit pipelining (coarse grain) involves predefining a sequence of frequently encountered primitive operations such as the well known multiplier | accumulator structure.

Pipelining techniques are aimed at improving the system throughput, usually at the instruction level. Since they depend on concurrent execution of different stages of a computation process, the improvement can be achieved only to the extent that the data dependency of the algorithm allows it. The instruction and intra-functional unit pipelining techniques also have the effect of shortening the clock cycle. These two techniques require only a moderate increase in control complexity and generally less resources than the inter-functional unit pipelining for which extra functional units must be added. But note that while the clock cycle is shorter for these techniques, the latency of a single instruction or operation will be increased because extra delays are introduced to the basic clock cycle due to the latching of intermediate results. Since the inserted latching delay cannot be partitioned, the gain in throughput will diminish while the latency continues to grow as the number of stages increases. Furthermore, if the pipeline resources are shared with other processes, then the pipeline must be flushed before switching between processes. When this happens, the idle time will increase as the number of stages increases. Therefore, the number of stages in a pipelined instruction or in an intra-functional unit design is usually limited to four or five. Even though the latency of an individual instruction or function does not decrease, a higher throughput rate in general can shorten the latency at the job level in varying degrees, depending on the data dependency inherent in the algorithm and the scheduling of the instruction execution.

Today's microprocessors all apply pipelining in the instruction execution designs. However, with the SIC approach, freedom in instruction/functional unit design is limited. Therefore, in most of the previous works, pipelining is mainly implemented at the inter-functional unit level. In a number of cases, even though the throughput and resource utilization are high, the speedup is unimpressive due to the lack of effective means to shorten the clock cycle. This indicates that the potential benefits of further exploiting pipelining techniques at lower levels, particularly with ASIC technology, will be substantial.

2.2.4.2.2 RISC

The concept of RISC has been a subject of both controversy and modification since its debut. It is necessary to distinguish the RISC design philosophy from its design techniques in order to obtain an objective assessment of its usefulness for future robotic computation hardware design. The fundamental principle of the RISC philosophy is to base the architecture resource allocation decisions on the analysis of the needs of the target applications and the potential benefits of a specific decision [GiMi87]. In contrast, other design philosophies may pursue goals such as support of system functions, support of a real-time environment, easy programming, or direct execution of high-level languages. Commonly cited RISC design features include a single instruction/execution cycle, load/store instruction set, fixed instruction format, hardwired control, a large register set, and a highly pipelined datapath [GiMi87].

If RISC is only an alternative design philosophy for general purpose computer systems, then it should rightfully be the default philosophy of any special architecture design. After all, if the architecture is designed for a special purpose, why shouldn't the allocation of the architecture's resources be derived from the analysis of its intended applications? Hence, for robotic applications, since the architecture design should be algorithm-specific, RISC is naturally the philosophy of choice.

In fact, the RISC concept has already been incorporated in a few recent works [WaBu87, JaOr87]. It has been previously pointed out that certain robotic control computations may require a throughput as high as 30 MFLOPS. Note that this figure is based on the sheer number of additions and multiplications needed without considering any parallelism. With the algorithm-specific design approach and the RISC feature of a single instruction execution cycle, a single chip run at 20 MHz can comfortably satisfy that computation requirement assuming a 3-stage FPM (Floating-Point Multiplier) in pipeline with an FPA (Floating-Point Adder) and a conservative speedup of 3. The hardware for implementing algorithms of this magnitude on a single chip or chip set is within reach of today's ASIC technology. Thus, it appears that most of the robotic control algorithms can be implemented on chips. From this perspective, the RISC concept is likely to play an important role in future robotic architecture design.

2.2.4.2.3 Systolic Array

A systolic array is a computation structure with the following features:

- Identical processing elements (PEs) are interconnected in a regular fashion;

- Each PE executes simple functions;

- Pipelining is implemented between PEs (and optionally within PEs) with lockstep synchronization;

- Data streams can flow in single or multiple directions.

Computation techniques based on systolic arrays have been successfully applied to image processing problems and matrix computations. However, algorithms that are to be efficiently executed on systolic arrays must have massive fine-grain parallelisms. While the formulation of the algorithm has some bearing on parallelism of this type, the eventual judgment is the degree of parallelism inherent in the algorithm and whether data movement can be achieved by local communication between PEs. The high-level formulations of some robotic algorithms (typically in matrix form) appear to be conducive to systolic implementation. But, the I/O is often a major communication bottleneck. In this case, neither speeding up the PE computation nor adding more PEs can increase the throughput or decrease the latency. Furthermore, even though systolic arrays have been designed to implement some basic matrix operations, such as inversion, the overall efficiency will still be low if only parts of an algorithm can be effectively executed in the systolic array structure. Because of the few design instances in previous work, whether systolic arrays can demonstrate certain advantages in executing robotic control algorithms is still an open question when comparing it with other computing structures.

2.2.4.2.4 Multiprocessors

For multiprocessor designs, the architect's main attention must be shifted from functional unit designs to the communication aspects of the system. Communication and the related control problems can be decomposed into three related issues of interconnection topology between processors and memory, communication protocol, and synchronization of data. Reviews of these issues can be found in [GaPe85, Kle85, GFCM87, AtSe88]. Among these issues, a crucial design decision in the context of the real-time constraint in robotic applications is the choice between shared memory and message passing.

Message passing is more conducive to computer networks and object-oriented programming environments. However, not only it is the more restrictive method of the two, but also it has the disadvantage of considerable overhead due to the protocol requirement. For example, for a process to read the value of a joint angle, another dummy process must be created (by the system) to respond to a request message from the reading process. The latency time for receiving the message back can easily go up to milliseconds [CFAB86]. This is several orders of magnitude greater than many of today's microprocessors basic operation time. This disadvantage seems to entail a severe performance penalty on multiprocessor systems, such as hypercubes, in which the interconnection between processors dictates that the communication protocol be based on message passing. In robotic control applications, processors are likely to be placed in adjacent locations, and the number of processors is usually small and will not pose a severe memory contention problem. Thus, with simple control such as priority access, interprocessor communication based on shared memory can approach the maximum speed of the bus and is thus preferred.

The parameter E_a is a useful measure particularly for indicating the matching of an algorithm with multiprocessor architectures. As an example, the E_a values computed from the performance data of the inverse dynamic computation for the Stanford arm taken from [KaNa85] are shown in Table 2-2. The number of processors used varies from one to seven, and in each case, the instruction scheduling is assumed optimal. It can be seen that E_a drops significantly as the number of processors exceeds 4. These results once again confirm the view that robotic control algorithms are characterized by moderate to strong serial dependency and do not have massive inherent parallelisms within a task to support cost-effective implementation on large multiprocessor systems.

Table 2-2. Speedup and Efficiency of a Microprocessor System
for Computing the Inverse Dynamic Equation.

Number of Processors	Processing Time (ms)*	Speedup S	Efficiency E_a
1	24.83		
2	12.42	2.0	1.0
3	8.43	2.945	0.98
4	6.59	3.768	0.94
5	5.86	4.237	0.847
6	5.73	4.3	0.72
7	5.69	4.36	0.623

* From [KaNa85].

More than half of the designs reviewed in this study are in the category of multiprocessors. Most of the systems designed for the upper layers are also in this category. Among the systems for the device interaction layer, most are homogeneous (using the same functional elements), but the most striking feature is the invariably *ad hoc* approach. The interconnect topology and control structure in most of these designs are mapped directly from algorithms. Memory organization and process communication are rarely specified. None of the architectures is designed for concurrent processing.

This situation seems to indicate that the traditional multiprocessor design paradigms are not quite sufficient for robotic architecture design. There are two main reasons. First, traditional multiprocessors are mainly concerned with very large scale scientific computations and widely different environments. But for the device interaction layer, especially the control domain, the computational need is relatively small and the application is fairly specific (the joint actuators). Second, for robotics control the algorithms are characterized by very large computational granularity, i.e., decoupled between tasks but high serial dependency within a task. Such characteristics result in poor hardware utilization when these algorithms are executed in architectures that aim at a large degree of parallelism.

As architecture designs for robotics are increasingly algorithm-specific and system designers are capable of using ASIC technology effectively, the emergence of another type of multiprocessor system is envisioned. This new type of multiprocessor system will be characterized by dedicated, algorithm-specific, single-chip processors cooperating to perform the device interaction functions. The distributed control methods, interface standards, and communication protocol designs of such systems for robotic applications will require additional research effort.

2.3 Robotic Kinematics

Manipulator kinematics is the study of all the geometrical and time-based properties of the manipulator motion without regard to the forces that cause it [Cra86]. Within the Device Interaction layer of the robotic computation hierarchy, the position and orientation of the manipulator are servo-controlled through each individual joint. At a higher level such as trajectory planning, kinematic information is more conveniently expressed in Cartesian coordinates. This creates the need to convert the kinematic information between the two coordinate spaces. The mapping from the joint space to Cartesian space is known as the Direct Kinematic Solution (DKS), and the inverse mapping, the Cartesian to joint space mapping, is called the Inverse Kinematic Solution (IKS) [Pau81, Cra86].

The complexity of robot control requires a hierarchical approach such that the entire problem is decomposed into self-contained modules with human-manageable complexity and a clearly defined interface [Ale*83]. At the Reasoning layer, a trajectory planner may accept commands from a higher level task algorithm and transform them into a sequence of path points. Each path point must be translated further through the IKS to joint angle set points and passed to the servo-control loop of the actuator. The IKS thus plays the central role in linking the high level algorithm into the lower level control mechanisms. In this section, the direct and inverse kinematic computations are first discussed, followed by a review of previous work on designing special architectures for computing the kinematic solutions.

2.3.1 The Direct Kinematic Solution

A robot manipulator consists of a number of nearly rigid links connected by joints, which allow relative movement of the neighboring links. The position of each rotational (or prismatic) link-joint pair is conveniently expressed by a single variable θ_i (or d_i) with respect to its own link coordinate system. A unique 4×4 homogeneous transformation matrix A_i, which is a function of θ_i (or d_i), maps a vector in the link i-th coordinate system to the link i–1-th coordinate system. Thus, for a six degrees-of-freedom robot arm, given its six rotational joint variables $\Theta = (\theta_1, \theta_2, \theta_3, \theta_4, \theta_5, \theta_6)$, the joint space to Cartesian space mapping is obtained by the successive multiplication of the six homogeneous transformation matrices,

$$T = A_1 \cdot A_2 \cdot A_3 \cdot A_4 \cdot A_5 \cdot A_6 = [\mathbf{n} \ \mathbf{s} \ \mathbf{a} \ \mathbf{p}]. \qquad (2.5)$$

The resultant homogeneous matrix T gives the orientation vectors of the wrist, $\mathbf{n}$, $\mathbf{s}$, and $\mathbf{a}$, and the current arm position, $\mathbf{p}$, all in the world coordinate system. The arm position $\mathbf{p}$ is defined as the vector from the origin of the base to the wrist.

2.3.2 The Inverse Kinematic Solution

More essential in the kinematic calculation is the IKS, which unfortunately has no unique solution in general as a given position for the robot arm end effector can be obtained from a number of arm configurations. Approaches for computing the IKS can be divided into two broad classes of numerical methods and closed form solutions. Numerical methods have the advantage of obtaining a general solution given that the structure of the manipulator is solvable, but because of their iterative nature, the computation is much slower than the corresponding closed form solution [Cra86]. A major result due to Pieper [Pie68] is that analytic or closed form IKS exists for a 6-DOF manipulator if three adjacent joint axes are revolute and intersect at a point with their twist angles equal to 0 or 90 degrees. For this reason, present day robots are designed with such characteristics. As a result, the closed form solution is chosen for implementation in this work.

2.3.2.1 Numerical Method

The numerical method for computing the IKS presented in this section is based on the modified predictor-corrector (MPC) technique by Gupta and

Kazerounian [GuKa85]. The basic approach in this method is to evaluate the joint angle values through integration of the joint rates by :

$$\dot{\Theta} = J^{-1}(\Theta) [\, \dot{x}_d + K \lambda\,] \qquad (2.6)$$

and

$$\Theta = \int \dot{\Theta}\, dt \qquad (2.7)$$

where Θ and $\dot{\Theta}$ are the vectors of the joint angles and their rates, respectively, and are of dimension $N \times 1$, (N is the DOF of the manipulator). J^{-1} is the inverse of a Jacobian, which maps the rate of change in Cartesian space to the joint space and is dependent on the current positions of the joint angles. $\dot{x}_d$ is the desired velocity vector along the trajectory. K is the gain matrix and the $K\lambda$ term is used to modify the end-effector rate.

In the usual predictor-corrector method, the predictor **P** is used only once to obtain an initial value of Θ for computing the joint rate $\dot{\Theta}$:

$$\Theta_1 = \mathbf{P}(\Theta_0,\ \Theta_{-1},\ \Theta_{-2},\ \Theta_{-3},\ \dot{\Theta}_0,\ \dot{\Theta}_{-1},\ \dot{\Theta}_{-2},\ \dot{\Theta}_{-3},\ \Delta t)$$

$$= 1.547652\Theta_0 - 1.867503\Theta_{-1} + 2.017204\Theta_{-2} - 0.697353\Theta_{-3} \qquad (2.8)$$

$$+ \Delta t[2.002247\dot{\Theta}_0 - 2.03169\dot{\Theta}_{-1} + 1.818609\dot{\Theta}_{-2} - 0.743200\dot{\Theta}_{-3}],$$

where Δt is the time step, and Θ and $\dot{\Theta}$ are the position and angular speed of the joint angles, respectively, with the subscript denoting the current (0) and past (negative) values. Once the estimated initial value of Θ is known, eq. (2.6) is evaluated, followed by the approximation of eq. (2.7) through the corrector **C** of eq. (2.9):

$$\Theta_1 = \mathbf{C}(\Theta_0,\ \dot{\Theta}_1,\ \dot{\Theta}_0,\ \dot{\Theta}_{-1},\ \dot{\Theta}_{-2},\ \Delta t)$$

$$= \Theta_0 + \Delta t[0.375\dot{\Theta}_1 + 0.791667\dot{\Theta}_0 - 0.208333\dot{\Theta}_{-1} + 0.041667\dot{\Theta}_{-2}]. \qquad (2.9)$$

The process of evaluation-correction is then repeated until Θ has converged or failed. The Δt can be used to dynamically control the performance of the algorithm.

Gupta and Kazerounian's modification is focused on the development of a scheme to determine the step size Δt based on an error measure of the difference between the desired and estimated position/orientation in Cartesian coordinates. Δt is determined as follows:

- Δt is reduced by half if the error due to the new estimate is either
  (1) greater than that of the current position and orientation, or
  (2) greater than the convergence criteria after 5 evaluation-correction iterations;

- Δt is doubled if the error of the new estimate due to the predictor satisfies the convergence criteria too well.

The modified predictor-corrector algorithm has been shown to be capable of finding an IKS if a solution exists or otherwise detecting the singularity case. This approach has the advantage of obtaining the joint rates simultaneously, but the computation takes several seconds and makes it impractical for real-time control.

2.3.2.2 Closed Form Solution

If a closed form solution is possible [Pie68], then the analytic solution for the joint angles can be obtained as follows. First, obtain the algebraic expressions for elements of U_6 to U_1, where $U_i = U_i \cdot U_{i+1} \cdot \cdot \cdot U_6$. Multiplying A_i^{-1} to the left hand side of eq. (2.5) recursively starting from $i = 1$ yields

$$T = U_1,\tag{2.10}$$

$$A_1^{-1} \cdot T = U_2,\tag{2.11}$$

$$A_2^{-1} \cdot A_1^{-1} \cdot T = U_3,\tag{2.12}$$

$$A_3^{-1} \cdot A_2^{-1} \cdot A_1^{-1} \cdot T = U_4,\tag{2.13}$$

$$A_4^{-1} \cdot A_3^{-1} \cdot A_2^{-1} \cdot A_1^{-1} \cdot T = U_5,\tag{2.14}$$

$$A_5^{-1} \cdot A_4^{-1} \cdot A_3^{-1} \cdot A_2^{-1} \cdot A_1^{-1} \cdot T = U_6.\tag{2.15}$$

If the above equations are solved from top to bottom, the left hand sides are always defined. Compare the matching elements of both sides of each equation and obtain an equation containing the sine and/or cosine of only one joint variable, the joint angle involved can then be solved. As an example based on the PUMA manipulator, the 14 and 24 elements of both hand sides of eq. (2.10) gives

$$P_x = \cos\theta_1 U_{214} + \sin\theta_1 d_3,\tag{2.16}$$

and

$$P_y = \sin\theta_1 U_{214} - \cos\theta_1 d_3.\tag{2.17}$$

After eliminating U_{214} via trigonometric substitutions, θ_1 can be obtained by

$$\theta_1 = \tan^{-1}(\frac{P_x}{P_y}) - \tan^{-1}\frac{d_3}{\pm\sqrt{r^2 - d_3^2}},\tag{2.18}$$

where $r^2 = p_x^2 + p_y^2$. The complete closed form solution for the PUMA is listed in Appendix A.

2.3.3 Previous Designs Dedicated to Kinematics Computations

The need to compute the kinematic solutions in real-time for intelligent control has motivated hardware implementations of DKS and IKS. A single VLSI chip architecture to implement the DKS computation has been proposed [LeSh87]. The design features fixed-point computation and on-chip calculation of trigonometric functions through table look-up and interpolation. Simulation results indicated that the DKS can be computed in 10 µs.

Gupta and Kazerounian's algorithm for IKS has been implemented in an architecture based on a commercial floating-point arithmetic unit [OrTs86]. To reduce the control complexity, the gain matrix K in eq. (2.6) is set to 0, the step size is fixed, and the magnitude is made small enough such that if the evaluation-correction result does not meet the convergence criteria, it automatically indicates the encounter of a singularity case. Simulation results indicated that the IKS for a 6-DOF manipulator can be obtained in 2,389 cycles for a single loop. Hence, with a clock rate of 10 MHz, the IKS can be computed in less than 0.25 ms.

An architecture has also been developed for computing the IKS closed form solution based on the cordic processor [LeCh87]. In this design, the closed form solution algorithm is first partitioned into tasks of roughly equal granularity executable on the cordic processor. A pipeline of 18 stages is constructed using 25 cordic processors to provide a throughput rate of one solution per cordic cycle. The number of delay buffers that are inserted between stages to synchronize the data flow is minimized via linear programming methods. With off-the-shelf cordic processors, the IKS can be obtained every 40 µs with an initial latency of 720 µs.

2.4 Summary

The impact of ASIC technology on system design has been examined. Issues concerning hardware implementation of robotic control algorithms have been discussed. Previous work on computer architecture designs for robotics have been surveyed. Background on kinematic computations have been presented.

The issue of how ASIC can be effectively applied to robotic control is addressed from two angles: the specificity and style of the architecture. Trends in robotic applications and the consequent market implications point in the direction of an algorithm-specific hardware design approach. Examination of previous robotic architecture designs reveals that the architecture styles of pipelining and RISC best match the characteristics of robotic control algorithms.

Chapter 3

A Conceptual Framework for ASIC Design

... [Research is] a strenuous and devoted attempt to force nature into the conceptual boxes supplied by professional education.

Thomas Kuhn

THE STRUCTURE OF SCIENTIFIC REVOLUTIONS (1970)

In this chapter, system design in ASIC is compared to that involved with SICs. The nature of ASIC design is then examined from both a transformation perspective and a decision making perspective. With this background, a conceptual framework for ASIC design is presented. This conceptual framework organizes the knowledge of IC system design into three knowledge frames: *design process*, *design hyperspace* and *design repertoire*. Key concepts presented in the process frame include the hierarchical approach, the role of methodology, and a model representing the implementation of methodologies. The hyperspace frame articulates the role of the design space concept and outlines the framing of the architecture and algorithm spaces as a means to facilitate recognition of design alternatives. The repertoire catalogs techniques for evaluating design alternatives. These three frames deal with different aspects of ASIC design, but they are integrated through an underlying theme of viewing design as a decision making process. That is, system designers must structure the design process so that the solution space is manageable and design alternatives are consciously sought and evaluated. Because of the growing importance of high-level design decisions, the discussion of these concepts will focus on one particular step — the transformation from task algorithm to architecture specification.

3.1 The Nature of ASIC Design

The term ASIC became popular only after 1985, but the semicustom design approach and all the basic ingredients of ASIC had already appeared before 1980. Interestingly, unlike the previous progress which lead to VLSI, it is difficult to single out a particular technological innovation, such as a new fabrication process, a memory chip, or a new microprocessor, that can symbolize the rise of ASIC. And yet this event is affecting a directional change in the discipline of system design. To gain more insight into this change, ASIC design and SIC design are first compared and the differences are inspected from two perspectives.

3.1.1 A Comparison of ASIC Design and SIC Design

ASIC design differs from the traditional SIC approach in a number of ways as summarized in Table 3-1. In microprocessor-based SIC designs, the performance is limited by the processor. The interconnections between various functional components usually do not pose a problem. In VLSI, however, interconnects consume up to 70% of the chip area and affect gate utilization. Also, they become the dominant factor in propagation delays as the level of integration increases.

Another important issue in the ASIC approach is the testability of the design. In SIC designs, nodes are accessible for testing, but they may become inaccessible in new ASIC chips. Testing currently takes 30-50% of the production cost of ASIC chips [Wal87]. System designers must incorporate a test strategy at the very beginning of the design cycle.

Table 3-1. A Comparison of Traditional Design and ASIC Design.

Attribute \ Approach	SIC Design	ASIC Design
Goal/Direction	chips to systems	systems to chips
Cost constraint	component count	design effort
Performance limitation	functional unit design	data communication
Major design alternatives	major components (e.g., processors)	design styles (e.g., GA, SC, FC)
Coupling between design steps	loose	tight
Testability requirement	nodes accessible at board level	must be incorporated early in the design
Verification	breadboarding	simulation
Prototyping	usually in-house	in cooperation with vendor
Last-minute changes	less costly	costly
Design guidance	informal	strong methodology
Tools	less CAE dependent	CAE intensive

3.1.2 A Decision-Making Perspective versus a Transformation Perspective

Design has long been regarded as a half-science, half-art discipline. This is probably due to the fact that design generally involves three different levels of activities as illustrated in Figure 3-1.

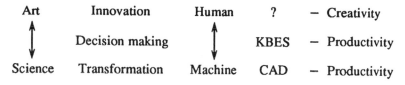

Art	Innovation	Human	?	— Creativity
	Decision making		KBES	— Productivity
Science	Transformation	Machine	CAD	— Productivity

Figure 3-1. The art-science dichotomy of design activities.

When viewing design as a creative process, hard-to-quantify factors such as individual talent, intuition, and experience, tend to dominate, moving the design activity closer to the human/art end. At the opposite end, transformation techniques derived from analyses can eventually be carried out by machines no matter how complicated they may be.

Corresponding to each level of design activity is a set of tools on which a whole generation of design modes evolve. At the bottom are the CAD programs that tackle various transformation tasks. Building on these programs, KBES (Knowledge-Based Expert Systems) are developed to facilitate the decision-making process. A shift from a design mode using CAD tools in a transformation perspective to a design mode using KBES in a decision-making perspective is currently underway. While the major objective of these two generations of tools is to raise productivity, they do not directly address creativity. One may speculate that the next generation of tools beyond KBES will address creativity, but currently there is little to suggest what forms such tools must take.

The transformation perspective tends to view design as a process of successive transformations of specifications from one domain (or abstraction level) to another. This is a powerful perspective so long as the function of transforming what to what is well defined. But as the design process migrates to a higher level, the specifications are prescribed according to the user's needs in terms of functional and performance requirements. These specifications are usually vague and incomplete; sometimes they may not even be feasible. In such a situation, the transformation perspective with a strong deterministic tone may not be adequate.

On the other hand, as a technology matures, major breakthroughs (such as novel processing technologies, device geometric structures, circuit techniques, or new system architectures) occur less frequently. At the same time, knowledge about interactions of a design and its environment continues to accumulate and will eventually be incorporated into CAD programs. As a result, design as a decision-making process — that is, the selection of a solution from a number of alternatives according to a set of cost/performance criteria — will dominate the design process. From this perspective, the two most important questions are: *what* design decisions are to be made and *how* are they made? In the remainder of this section, decisions that need to be made in ASIC designs are discussed. The conceptual framework basically addresses the issue of how these decisions are to be made.

Design decisions can be classified into four categories:

- Software/hardware tradeoffs;
- Processing technology;
- Implementation style;
- Choice of hardware algorithms.

Decisions on software/hardware tradeoffs affect the flexibility of the product and thus are mainly influenced by the need to modify the design in the future. Hartmann has provided an excellent discussion of this subject [Har86]. The merit of a particular technology is usually judged by its gate delay, power consumption, noise immunity and logic capacity. A comparison of various technologies can be found in [Sol82], while the reasons for the dominance of CMOS as today's chosen technology are explained in [Che86, Sho88]. Implementation style choices include programmable logic devices, gate arrays, standard cells, and module generators. Economic tradeoffs between these design styles have been studied in [FePa87, EBCH86].

Hardware algorithms, as distinguished from high-level task algorithms, refer to functional module designs in multipliers, ROMs, RAMs, and PLAs where regularity in the structure can be captured and exploited in a procedural fashion [Yas86]. Note that the first three kinds of design decisions are largely influenced by marketing considerations such as design standards, compatibility requirements, expandability, product lifetime, and other economic factors. In most cases, they can be determined before the actual design activity starts. In contrast, even though the choice of hardware algorithms greatly affects the quality of a design, it may be necessary to delay this design decision due to the uncertainties in meeting physical constraints such as chip area, power consumption, I/O bandwidth, or system partitioning. Furthermore, even when tradeoffs among alternatives are clear, decisions must be made in context. For example, the tradeoff between the size and delay time of a transistor is well understood, but the decision of whether time or area is to be optimized may depend on whether the transistor is in the critical delay path of a circuit. It is this dynamic nature of decision-making and the combinatorial explosion of alternatives, composed mainly of hardware algorithm choices for various functional units at various levels, that constitutes the fundamental problem in VLSI design.

3.2 The ASIC Design Process

The fundamental attribute that governs many aspects of the VLSI design process is the unprecedented complexity, which has arisen from the combination of the tightly-coupled nature of numerous intermediate steps and the seemingly unlimited freedom of design choices at each step. As a result, the design methodology as well as its implementation — the CAE tools and the environment in which these tools are integrated — characterizes the VLSI design process.

3.2.1 VLSI Design Hierarchy

To combat complexity, the strategy of "divide-and-conquer" must be employed. This approach is manifested as a hierarchy of design steps. In essence, the hierarchical approach partitions various aspects of VLSI circuits into abstraction levels and defines the order among these levels. A methodology is then a particular ordered sequence of steps linking these abstraction levels. A set of CAE design tools is required for the implementation of the methodology. Table 3-2 illustrates the generally accepted hierarchical levels, the abstractions they represent, and the supporting CAE tools they require.

3.2.2 VLSI Design Methodology

VLSI design methodology is a formalization of the VLSI design process. At the beginning of the evolution of the design methodology, answers to the questions of what to abstract and how to abstract were largely guided by individual designers. They are inevitably constrained both by the available resources and the maturity of the knowledge base. The subsequent partitioning process and the ordering of the abstraction levels either follow a natural style or are done in an *ad hoc* manner. CAE tools were first developed individually for each design task and later combined to form more powerful automated systems. While this *ad hoc* style of design methodology implementation is efficient and may even be necessary at the early stages of development, it is recognized that the side effects of such an approach are becoming an unbearable burden for the development of integrated design systems. For example, incompatible data formats require translation programs, which not only become an overhead to the system performance but also complicate the data management effort. Also, the lack of any industry standard is at least partially responsible for the still relatively high development cost and the slow proliferation rate of these systems. The moral seems to be that the success of a hierarchical design methodology depends, on one hand, on how well its underlying principles reflect the nature of VLSI circuitry and, on the other hand, on how well its external expression supports the implementation effort. These can informally be called the necessary and sufficient conditions for efficient implementation of design methodologies.

3.2.3 DOEMA: A Model for Methodology Implementation

A unified conceptual model of the design process is indispensable not only for the efficient implementation of the hierarchical design methodology, but also for the user to master the corresponding CAE tools with as little effort as possible. The model should be:

Table 3-2. Hierarchy Levels, Abstractions, and Supporting Tools.

Hierarchy Level	Abstraction (Aspects Represented)	Supporting Tools
System	space-time behavior as instruction, timing and pin assignment specifications	flow charts, diagrams high-level languages
Architecture	global organization of functional entities	HDLs, floor-planning, block diagrams, programs that estimate areas and clock cycles
Register transfer	binding of data flow functional modules, microinstructions	synthesis, simulation, verification, and test analysis programs; programs for evaluating resource utilizations
Functional modules	primitive operations and control methods	libraries, module generators, schematic entry, test generation programs
Logic	Boolean function of gate circuits	schematic entry, synthesis programs, simulation and verification programs, PLA tools
Switch	electrical properties of transistor circuits	RC extraction programs, timing verification and electrical analysis programs
Layout	geometric constraints	layout editor/compactor, netlist extractor, design rule checker, floor-planning, placement and routing programs

- Concise, consistent, and complete;
- Capable of utilizing existing CAE tools;
- Capable of facilitating CAE development efforts;
- Capable of accommodating future changes;
- Distinctive in the definition of roles for human and machine;
- Capable of accommodating human intervention to observe and fine-tune;
- Easy for users to learn;
- Representative of aspects common to various abstraction levels;
- Separate in mechanism and content.

A conceptual view of the design process, so formulated, is illustrated in Figure 3-2. From a system point of view, this model can be described in terms of *Design Object*, *Design Engine*, *System Manager* and *Expert Assistant* (**DOEMA**). Each of these components will be discussed in the following sections. Some general aspects of this model are first discussed.

The fundamental characteristic of this design process model is its handling of level-specific design process information. Similar to Walker's thesis-antithesis-synthesis argument of AI [Wal86], this model can be viewed as a synthesis of the thesis and antithesis of design methodology. The thesis is that a

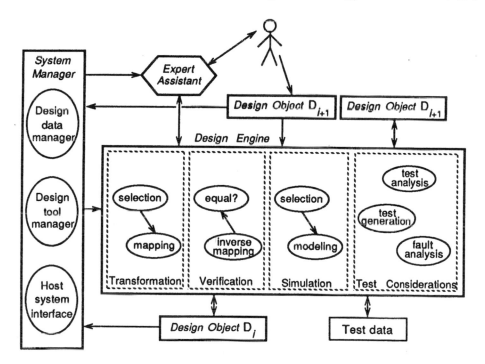

Figure 3-2. The DOEMA model for implementing ASIC
design methodologies.

model of translating specifications from a higher level to a lower level, general to all levels, is possible [Lip83]. However, because of the wide differences between the subjects of various abstraction levels, the necessary inclusion of all these differences in a general model makes its execution very inefficient. The antithesis is that since the subject of abstraction in each level is so different, a design process model for each translation step should be developed. The execution is more efficient but it now exhibits a higher development cost. Note that a silicon compiler can be built from either of these design process perspectives. The synthesis here is essentially taking a middle ground of the two approaches. It retains the general model idea of the thesis, but instead of putting everything in this model, it separates the level-specific information from the subprocesses that are common to all levels.

To represent the design process in terms of an inter-level model is inherently simple, consistent, and without loss of generality. From the CAE tool developer's point of view a single "design engine" can be built since a general design process model is uniformly applied to all levels. From the user's point of view, the model is simple with built-in consistency; learning efforts are thus minimized. In addition, the open-architecture style of the DOEMA model allows logic synthesis systems to be built from existing modules or modules from different vendors. Concurrent design activities at different levels are also possible.

3.2.3.1 The Design Object

The design object of level i is a description of a target design in terms of the abstractions defined at that level. It may be viewed as a collection of abstract objects and is defined as a point in the level i design space. For example, a 16-bit multiplier is an instantiation of an abstract object called n-bit multiplier at the functional module level. To describe the object, *languages* are necessary. Primitives and constructs provided by a language, however, are based on an evolving understanding of the design space. By using a language to describe an object, the level i properties of the object are thus transformed into the three basic attributes imposed by the language itself: the description (data) format, the syntax, and the semantics. In the case of the multiplier, for example, the logical relationship between the bits (from the most significant to the least significant) is encoded into the data format (a particular order in the numbering of the bits), and the algorithmic/structural property (such as shift-and-add or array multiply) is embodied in the semantics, all conforming to the syntax stipulated by the specific language used. Through these three attributes, human and machine can communicate.

The concept of design object plays two important roles in the DOEMA model. First, it is the place where the dynamic (mechanism) and the static (knowledge about the design space) aspects of VLSI design intersect. By limiting the intersection to within the boundary of the design object, the remaining dynamic aspects can be separated as purely mechanistic, which in turn allows the design engine to be developed. Second, it is also the place where two adjacent levels within the hierarchy interface. Once the data format, syntax, and semantics for each level are defined, the coupling of the levels becomes straightforward.

3.2.3.2 The Design Engine

The design engine represents the mundane tasks of VLSI design, and evolves from the four design subprocesses: *transformation*, *verification*, *simulation*, and *test consideration*. They are uniformly present at all design levels. The current state of implementation is obviously very level-specific. But as a conceptual model, the level-specific information is best kept separated so that the essential aspects of the design dynamics remain the focal point.

The Transformation Process

The transformation process is the most fundamental among the four subprocesses. This process can be further partitioned into two mechanisms: mapping and selection. The difference between the two is whether or not high-level decisions are involved. Mapping represents purely mechanistic and routine procedures (e.g., a transistor is formed by a diffusion region crossed by a polysilicon path) or involving only low-level decisions that can be parameterized (e.g., the aspect ratio of a transistor can be related to the transition time). Selection, on the other hand, represents the most important decision-making function of the designer or the design system (they are interchangeable in many cases) — to select a particular circuit implementation from a pool of alternatives given the area-time tradeoff. For example, a desired logic function can be realized by a dynamic logic design instead of a conventional static design. Or, to improve the testability of the circuit, well known structural techniques such as scan-path methods or built-in self test (BIST) can be chosen [WiPa83, McC86]. The distinction of the mapping and selection mechanisms emphasizes the human role in evaluating design alternatives and making design decisions.

The Verification Process

Verification is the process of confirming that the resultant lower level design object is indeed the intended design. This process should be distinguished from validation, which simply checks for violations of various design rules. In practice, verification is often done by simulation, simply because a formal approach to prove the correctness of a design either does not exist or is impractical [WiPa83]. Verification in this model is achieved by first generating another level $i+1$ design object through inverse mapping and then by an equivalence test of the two level $i+1$ object descriptions. For example, CVS (Connectivity Verification System) was one of the earliest systems built reflecting this approach [NeVi86]. After the transistor-level design object has been transformed to a corresponding design object at the layout level, CVS extracts from the layout description another transistor-level design object. This second version is compared with the original design object for connectivity equivalence. Similar procedures can be carried out at other well understood levels such as logic-gate or gate-transistor levels [Jac86]. The major advantage of this approach is its potential for tremendous savings in simulation time. Time savings of orders of magnitudes at certain levels have been reported [NeVi86]. The feasibility of this approach clearly depends on the existence of the inverse mapping and a canonical representation of a design object. There are some indications that nascent AI techniques, such as theorem proving and automated reasoning, may play an important role in this area [KaWo85].

The Simulation Process

The simulation process involves two steps: first, the generation of an appropriate model for the design object according to the level at which it is described; and second, an analysis of the output and next states given a set of input and states. The simulation process serves three different purposes: verification, optimization, and fault coverage indication. Whereas inverse mapping is not well developed, verification is achieved by extensive simulation. A selection mechanism within this process determines the appropriate circuit models to be used. It also determines whether global or local simulation is performed. Since simulation is based on the behavior of the lower level models, the result must be extrapolated back to the higher levels, a process that is now mostly left to the designer.

Testability Considerations

A unique characteristic in VLSI design is the role of design for testability [WiPa83]. Without the upfront objective of DFT, testing of custom or semicustom circuits of high complexity is virtually impossible. Test considerations include testability analysis, test vector generation, and fault coverage estimation [SeAg85, HnWi85]. Test analysis programs evaluate the testability of a design based on how difficult the values of the internal circuit nodes can be set (controllability) or retrieved (observability). Test generation programs attempt to identify and reduce input patterns that will cause as many faults as possible to be detectable at the outputs. The effectiveness of the test generation program is measured by the size of the resultant input test vector set and the percentage of faults it can detect (coverage). Fault coverage is usually approximated through statistical estimates on random test input vectors. More expensive exact coverage can be obtained by systematically injecting faults into the circuit and analyzing the simulated outputs [Hui88].

3.2.3.3 The System Manager

The role of the system manager is to provide an integrated environment in which the engine operates. This includes design tools management, design data management, and host system interface.

In the DOEMA model, the tasks represented by the design engine are generic and they must be instantiated for use at a specific level. For example, the program that transforms a gate level design into a switch level design is very much different from that between the switch level and the layout level. In addition to this instantiation function, the design tools management must provide structure and environment that can integrate tools from different vendors and allow incremental changes. Ideally, this latter service provides the DOEMA model some flexibility in accommodating the diversity of existing CAD tools.

Design data management is more than routine database management. It must support hierarchical representations of the design objects, alternative implementations, and evolutionary versions [Kat85, BrGr87]. It must also provide control mechanisms for secure and concurrent access and for recovery from system failures. Host system interface includes a common set of primitive functions such as file system support, process control, and communication protocol. With this interface to the host operating system, the model can be implemented independent of hardware platforms.

3.2.3.4 The Expert Assistant

Technology breakthroughs, creative innovations, and efficient management can all contribute to the improvement of design quality. While the former two often bring about revolutionary effects, they are rare and thus are somewhat out of direct control. Management, in contrast, is something on which one can always have a tight grip. In fact, the sole purpose of design methodology is aimed exactly at this — the management of complexity.

VLSI design complexity is a direct result of the large degree of freedom in placing and interconnecting devices on silicon and, thus, the numerous design alternatives for a given problem. As invention of new circuit techniques occurs less frequently while knowledge continues to accumulate, for a majority of system architects the major function as a designer increasingly shifts toward evaluating existing alternatives rather than creating new ones. An awareness of various alternatives is thus the precondition for making judicious design decisions. This intrinsically depends on the system designer's knowledge of VLSI technology.

To help the designer master the knowledge, nascent expert system technology based on AI research may be the key. Contemporary AI thinking perceives that knowledge is of two kinds: domain knowledge consisting of general rules and specific facts; and meta-knowledge, the knowledge about the application of rules and the reasoning mechanism. From this view, expertise is the act of applying domain knowledge to specific problems and thus can be mechanized once these rules and facts are clearly defined. Thus, expert systems are constructed with three components: *working memory* (WM) for representing the current state of the problem at hand; a *knowledge base* containing a set of production rules of the form "IF <condition in WM> THEN <action>"; and an *inference engine* that carries out the reasoning mechanism under a control strategy to determine the next rule to be activated [FiFi87]. The separation of the domain knowledge (representation) from the general reasoning mechanism is the essential feat on which the success of the expert system technology is based. By encoding knowledge in the form of rules, expertise can be captured and reproduced in machines.

In the DOEMA model, the major function of the expert assistant (a KBES program) is to facilitate the decision-making process by making the designer aware of design alternatives. To achieve this goal, the expert assistant requires two kinds of domain knowledge: one kind for determining when alternatives should be considered, and a second kind for determining what the alternatives are. The former can be captured in the design schema, which serves as a roadmap or overall procedural guide for conducting the entire design process. Alternatives are encoded as constraint-criteria templates, which list all possible actions in various situations based on previous experience. Such systems are currently being developed by ASIC vendors [Ano87].

3.3 The ASIC Design Hyperspace

While the dynamic aspects of VLSI design are represented by the design process model, the static aspects are best captured in the notion of a design hyperspace. Static aspects refer to properties that are inherent in the design objects and are largely stable over time. The significance of the design space concept lies in its classification power, which allows the designer to accumulate and retrieve knowledge efficiently and thus facilitates the recognition of alternatives. The relationship between the design process and the design hyperspace can be represented by Gajski and Kuhn's Y-chart as shown in Figure 3-3, in which the design process is formulated as a methodology traversing through the design hyperspace in a spiral fashion with more detailed information added toward the center [BuMa85, GaKu83]. However, the Y-chart does not address the details of individual levels. And, because of the multi-faceted nature of VLSI design, the framing of the design hyperspace is not unique; much depends on the designer's perspective. In order to frame it properly, an examination of the space concept is worthwhile.

3.3.1 The Design Space Concept

The space concept is heavily utilized in mathematics as a generalization instrument to study properties of abstract concepts. Two observations can be made on how mathematicians create the frame of reference for the space under investigation. One observation is that there are essentially an infinite number of frames one can create for a space. For example, for a 3D space we have Cartesian, cylindrical, or spherical coordinate frames. While these are all valid frames, differences exist in the degree of convenience encountered in using a particular frame to characterize an object (in this case, a mathematical function) under certain conditions. It is obvious that the choice of a particular frame is heavily influenced by application and guided by experience. On the other hand, no matter which frame is chosen, they all possess the one fundamental attribute of *orthogonality*, or mutual independence of the coordinates. This idea is

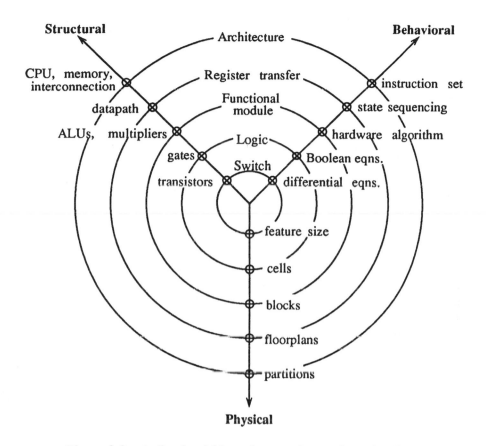

Structural

Behavioral

Architecture

CPU, memory, interconnection

Register transfer

instruction set

Functional module

datapath

state sequencing

ALUs, multipliers

hardware algorithm

Logic

gates

Boolean eqns.

Switch

transistors

differential eqns.

feature size

cells

blocks

floorplans

partitions

Physical

Figure 3-3. A five-level hierarchy superimposed on the Y-chart representation of the VLSI design hyperspace.

particularly useful in framing the design hyperspace, for, if the framing is orthogonal, then the effort of managing the combinatorial complexity of the design alternatives can be reduced by limiting our attention to only one dimension at a time.

With this in mind, the concept of the design hyperspace can be explored. The purpose is twofold: first, to frame the design hyperspace so that the designer can more readily recognize possible design alternatives. Second, some assurance that the framing can facilitate the mapping of the design objects is desirable. To achieve these goals, a multi-level view of the design hyperspace as advocated by Dasgupta and others is adopted [Das84, Tan84]. An overview of the design hyperspace is shown in Figure 3-4. Pertinent to the subject of this thesis, the algorithm space and the architecture space are of particular interest. Furthermore, since the final target is a VLSI architecture and the goal is to organize alternatives, the investigation begins with the architecture space and the result is then used to guide the study of the algorithm space.

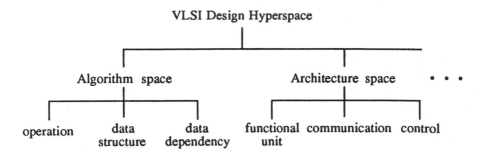

Figure 3-4. An overview of the VLSI design hyperspace.

3.3.2 The Architecture Space

The framing of the computer architecture space has been attracting attention since Bell and Newell's PMS (Processor/Memory/Switch) notation [BeNe71]. Despain recently presented a more modern view of the architecture space as shown in Figure 3-5 [Des84]. A computer architecture can be characterized by its *control concurrency, data specification,* and *data in-statement.* The control concurrency dimension specifies how the atomic calculations are done. The data specification dimension represents various choices for specifying where data values are located. The data in-statement dimension refers to the conditions under which data values may be changed. This framing scheme presents a highly abstract view of architecture in the sense that it is concerned with the logical aspects of a computational machine only. Even though such classification has the advantage of covering a wide range of

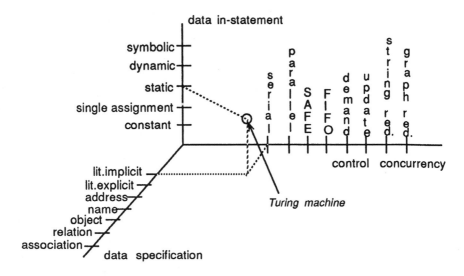

Figure 3-5. Abstract architecture model space [Des84].

machines, it may be overkill in the ASIC environment where finding a suitable architecture to implement the task algorithm in a single chip or a chip set is the major concern.

In view of this, a composite view of the architecture space slanted toward a hardware implementation perspective is presented. From a highly abstract view, information that is ultimately represented in the form of electrical signals (charges) in the hardware is processed in one of two ways — *transformation* or *transportation*. To make this fact explicit, the architecture space is viewed as being composed of three subspaces: *functional unit, communication,* and *control.* The first two subspaces represent resources corresponding to the transformation and transportation, respectively. The control subspace provides resources for binding them.

The Functional Unit Subspace

The functional unit subspace is spanned by three dimensions: arithmetic, logical and storage. Arithmetic units include conventional fixed-point and floating-point adders, multipliers and the like, as well as specially designed units such as a cordic unit. Logical units include shifters, multiplexers, decoders, comparators, and other modules for logical operations. Storage units include latches, flip-flops, registers, and memories. Storage units can be made to transform information in a limited way; their major role, however, can be viewed as an interface between transformation and transportation. With this framing scheme, queues and stacks, for example, can be visualized as a point in the functional unit subspace indexed by shifters and registers.

The Communication Subspace

The communication subspace contains the dimensions of interconnect topology, memory organization, and data reference. Major interconnect topologies for data exchange networks include direct links, bus, ring, star, tree, crossbar, mesh, and shuffle networks. Memory organization includes cache, local, and global structures. Data reference refers to various memory addressing schemes.

The Control Subspace

The control subspace has the dimensions of synchronization, style, and structure. The synchronization dimension consists of synchronous and asynchronous control, each elaborated by clocking schemes and protocols. The control is also characterized by the style of instruction execution such as centralized, pipelined, and/or distributed. The structure dimension includes implementation structures of random logic, PLA, and microcoding.

When the architecture space is framed in this way, different architectural primitives with similar high-level appearances naturally tend to cluster close to each other. Implementation alternatives for a given algorithm are thus more visible.

3.3.3 The Algorithm Space

Even though considerable research effort has been devoted to characterizing parallelism in programs and algorithms for matching them to languages and architectures, a better understanding is still being sought [JaGD87]. For the purpose of this work, it is desirable to have a frame of reference for the algorithm space that has a direct correspondence with the frame for the architecture space. Kung's characterization of the algorithm space provides a starting point.

To characterize parallel algorithms, Kung proposed a three dimensional space spanned by the dimensions of *concurrency control, module granularity,* and *communication geometry* [Kun80]. Concurrency control refers to the synchronization methods for correct execution of parallel algorithms. Module granularity refers to the computational size of a task module and is hardware dependent. Communication geometry includes various interconnection patterns for data communication between functional modules. Similar to this characterization, we view the algorithm space as being composed of three subspaces: *operation, data structure,* and *dependency*.

The Operation Subspace

The operation subspace is spanned by elementary computations seen by the designer. What the designer sees depends on the availability and his/her knowledge of hardware primitives. This subjective characteristic is inevitable, because a task algorithm can indeed be realized in more than one way, thus having more than one image in the operation subspace. Consequently, it matters as much for the designer to determine how "elementary" (i.e., the granularity) the computation primitives should be as to realize hardware alternatives. A primitive operation represents an event occurring in some entity or functional unit. Thus, the operation subspace can naturally be aligned to the functional unit subspace. This strict correspondence between operation and functional unit subspaces allows the algorithm designer to take the viewpoint of an architect in considering hardware alternatives.

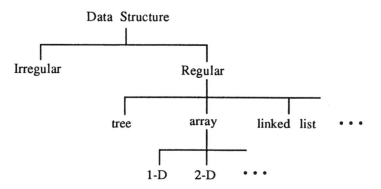

Figure 3-6. The data structure subspace.

The Data Structure Subspace

Data structure is the physical or visible form of the structural properties intrinsic to the task algorithm. The scope of this subspace is similar to the communication geometry described by Kung and is illustrated in Figure 3-6. Data structure has a major impact in shaping the proper organization of communication resources of an architecture.

The Data Dependency Subspace

"Characterizing the data dependencies lies at the heart of the algorithm-to-architecture mapping problem" [Jam87]. Padua, Kuck, and Lawrie have identified four kinds of program dependencies, which constitute the dependency subspace [PaKL80]. They are flow dependence, anti-dependence, output dependence, and control dependence as illustrated in Table 3-3. Regular control flow and data flow patterns, such as array operations, looping, and recurrence, can be expressed in terms of these dependencies. Mapping dependency patterns to architectures has been intensively studied by others and is reviewed in [JaGD87].

Algorithms at the task level are invariably developed with specific perspectives. The abstraction mechanisms or techniques at this level are primarily determined by the nature of the problem at hand. For example, a control engineer may be concerned with the eigenvalues or controllability of a matrix without considering how the actual computations are carried out.

Table 3-3. Fundamental Data Dependencies.

Dependency Type	Algorithmic Form	Symbol
Flow	S_j: $\quad x := op(...);$ $\vdots$ S_k: $\quad \_ := op(...\, x\, ...);$	$S_j \downarrow S_k$
Anti	S_j: $\quad \_ := op(...\, x\, ...);$ $\vdots$ S_k: $\quad x := op(...);$	$S_j \not\downarrow S_k$
Output	S_j: $\quad x := op(...);$ $\vdots$ S_k: $\quad x := op(...);$	$S_j \Downarrow S_k$
Control	S_j: $\quad x := op(...);$ $\vdots$ S_k: $\quad$ if (x) then ... ;	$S_j \dashrightarrow S_k$

However, once the algorithm has been developed and hardware implementation is considered, the designer must adjust his/her perspective to that of the computer architect. The foregoing framing scheme of the algorithm space has the advantage that each of the three architectural resources has a major, though not exclusive, correspondence with one particular algorithm subspace, namely, functional units with operation, communication with data structure, and control with data dependency. As a result, the designer's adjustment of his/her perspective to that of the computer architect is facilitated.

3.4 The ASIC Design Repertoire

The previous two sections have established the two basic concepts of design, namely, the *process*, as formalized into methodology and implemented by CAE systems, and the *hyperspace*, as the collective logical and physical properties of design objects. In this section, the actual mapping is considered in more detail, with particular interest focused on the architecture level.

Setting aside the mechanistic mapping of design objects, design becomes a decision making practice. Decision implies alternatives. Through methodology the search space is limited and hence the number of alternatives. Through the framing of the design spaces their existence is recognized. Then, the logical question to follow is: how should these alternatives be evaluated? The design repertoire is a collection of analysis techniques for evaluating design alternatives. It enables knowledge of designs techniques to be accumulated in a more organized manner and thus minimize possible redundant engineering efforts.

Research efforts aimed at understanding how the algorithm to architecture mapping should be done are conducted along two basic lines. The first is to find a suitable architecture for a given algorithm. The second is to restructure a given algorithm to better match a given architecture. These two approaches are termed *resource configuration* (RC) and *algorithm restructuring* (AR). In addition, *system partitioning* techniques have been developed to improve the productivity of implementation and to meet physical constraints such as area and I/O pin limitations.

The interaction between algorithms and architecture is not a one way relationship. Even though we distinguish the RC and AR approaches for classification purposes, no single approach is adequate. In fact, automatic synthesis programs based on these techniques typically combine these approaches in various ways [Thet83, Tho86]. The following sections present the current understanding of these techniques.

3.4.1 Resource Configuration

The objective of this technique is to determine an appropriate architecture for a given (fixed) algorithm. In other words, it is to determine the minimum resources needed to satisfy some cost-performance criteria. From the perspective of the architecture space, there are three kinds of resources as explained before: functional units, communication, and control.

Graph algorithms are often used for the configuration of functional resources [Des84, PaKL80, Kuc77, TsSi86]. The basic idea is to model the elementary computations as nodes and the data dependencies as arcs connecting relevant nodes. Then the parallelism or the data dependency can be derived from the upper bound on the size or depth of the equivalent tree. To derive a minimum resource solution, the graph is partitioned into cliques (or clusters), which correspond to various resources, and serial-to-parallel transformation is then applied to generate various alternatives for evaluation.

Communication includes interconnect topology and memory organization/addressing schemes. Conventional bus structure is well understood with the choice of a particular bus made according to the application [DaDo86]. Memory bandwidth constitutes the ultimate performance limit that an architecture of this type can attain since memory and I/O requirements must be satisfied under these conditions. The disparity between on-chip and off-chip delay in VLSI favors local communication. The application of special interconnection networks such as the butterfly connection, the shuffle network, etc., is often associated with well known domain-specific characteristics of data structure and data flow patterns, as can be seen in the areas of signal processing and sorting. Memory structure is crucial to relieve the communication burden, particularly in array-related operations. Discussions on various memory organization and addressing techniques can be found in [HwBr84].

Control plays the central role of binding the functional units with the communication facilities. Configuration of this resource depends on the application and on decisions regarding the other two resources. Kung has documented control patterns for three categories of applications (signal processing, matrix arithmetic, and non-numeric applications) in systolic array designs [Kun80, Kun84]. For computation-bound algorithms, pipelining techniques may improve throughput with modest increases in hardware. Theoretical bounds on the complexity of an algorithm for synchronous pipelined processing have been studied in [ScAt77]. As to implementation styles, control logic is increasingly implemented in PLA as sophisticated minimization and area optimization techniques are quite well developed [NeVi87].

3.4.2 Algorithm Restructuring

The objective of AR techniques is to modify either the data structure or the flow pattern of a given algorithm so that it can be efficiently implemented on a fixed computer architecture. AR techniques for microprocessor-based designs have previously been implemented in the Harpy system to evaluate architecture alternatives [BiMR83]. Since the constraints of the ASIC design environment are significantly different from the microprocessor environment, further development of these techniques is needed. In particular, the regularity of an algorithm takes on a new dimension of importance.

While all AR techniques are aimed at modifying the data flow patterns within the constraints of the function and data dependency of an algorithm, the emphasis varies. For algorithms exhibiting massive parallelism, as in array-type operations, the emphasis is the data structure and data communication (including processor-to-processor and processor-to-memory). It has been realized that they are the most critical ingredients in the performance of an algorithm under a given architecture [GaRo84]. Therefore, restructuring techniques are aimed at either improving the regularity of the data structure or reducing the data communication. In some situations, dramatic results can be obtained by simply applying basic elementary operations under the constraint of data dependency. More often, AR techniques for this class of algorithms involve partitioning of the array operands. Especially in pipeline architecture or systolic arrays, global data communication is reduced or transformed to local communication. This is achieved by eliminating the need to store intermediate results by partitioning and scheduling of the operations based on the characteristics of the assumed architecture [GaRo84, NiHw85, Jaet85, Ary85].

For algorithms that do not exhibit parallelism, freedom in restructuring is limited to better allocation of resources and scheduling of operations. In this case, algorithms are represented in data flow graphs and algorithmic or heuristic methods can be applied to determine optimal allocation and scheduling [PaHa87, Tho86].

3.4.3 System Partitioning

The partitioning of a large system into smaller modules in order to satisfy physical constraints has long been a concern in PCB designs. For ASICs, while the increasing number of gates that can be integrated on a single chip will lessen the importance of partitioning, higher integration usually costs more, and there are always some restrictions on the IC packaging capabilities. More importantly, subsystem sizes, as the result of partitioning, have a direct impact on productivity. It has been demonstrated that design activity is most efficient if the design manpower requirement is kept under 29 manweeks [FePa87]. As a result of better productivity and possibly larger parallel design efforts, scheduling time can be significantly reduced.

The partitioning of the system in the PCB design environment is generally aimed at minimizing one or more of the following:

- number of clusters (blocks of circuits that are closely related);
- circuit delays;
- cluster interconnections;
- variance of cluster sizes.

Semiempirical techniques based on what is known as *Rent's rule* have been developed to relate the average number of pins per module to the average number of blocks per module [Hol87, LaRu71].

While techniques previously developed for SIC designs are useful in ASIC design, different constraints (silicon area-delay time in ASIC vs. board area-delay time in PCB) mean that it may not be advantageous to have the system partitioned into chips in the same way that it was partitioned into PCBs. For example, circuit blocks previously distributed on more than one PCB may now be more efficiently integrated into a single chip such that the off-chip I/O traffic is reduced. Furthermore, recent study has shown that partitioning of the system along a functional line can result in a smaller number of pin-outs and reduced interconnection length [Fer85]. Some projects have been aimed at making the partitioning problem more tractable. For example, Palesko and Akers presented an algorithm for logic partitioning [PaAk83]. But the application of their method is limited to gate arrays and the partitioning problem as a whole is still much dependent on the designer's experience and intuition.

3.5 Summary

A conceptual framework for ASIC design has been described. It is composed of the three knowledge frames of *design process*, *design hyperspace*, and *design repertoire*. They address different aspects of how design decisions are made. The global strategy is to limit the search space through methodology implemented by CAE and KBES. Alternatives are recognized through a proper framing of the design space.

The conceptual framework equips the system-IC designer with a clear and coherent view of VLSI design activity and, at the same time, suggests a systematic way to acquire and accumulate knowledge. In this respect, it provides new goals for engineering education of future IC designers. This conceptual framework also gives CAE developers a unified and farseeing view to implement design methodologies in an integrated environment. Common understanding developed between tool developers and IC designers as a result of this work can speed up the vital transfer of ASIC technology to a much wider engineering community.

Chapter 4

The IKS Chip Design Paradigm

Rules, I suggest, derive from paradigms, but paradigms can guide research even in the absence of rules.

Thomas Kuhn

THE STRUCTURE OF SCIENTIFIC REVOLUTIONS (1970)

[A]n example equals one thousand inferences.

Hubert Dreyfus and Stuart Dreyfus

MIND OVER MACHINES (1988)

4.1 Introduction

As is true in any realistic project, the design of the IKS chip has gone through numerous refinements during its course. Desirable as it may be, documenting all the changes is itself a formidable task. The multi-dimensional and iterative nature of the design activity make it difficult to describe the design process concisely in an essentially linear text. But even if it can be done, documentation alone is not a substitute for understanding, which is what this work sets out to accomplish in the development of a paradigm. To achieve this goal, various design decisions that lead to the final design are examined and organized into the form of an ASIC architecture design methodology. The design of the IKS chip is then presented as an execution of this methodology. To better focus on design decisions and to avoid over-complexities, the presentation of the structural decomposition of the IKS chip is delayed until the next chapter where it is presented together with the VHDL simulation results. Testability issues are addressed in each design phase when appropriate and are assessed qualitatively as part of the evaluation of the overall design. In the remainder of this section, the assumptions, objective, and philosophy of this design effort are first examined.

4.1.1 Assumptions and Constraints

The IKS chip design is based on certain assumptions about its intended use and the implementing technology. These assumptions reflect a compromise between an ideal design environment and a practical one with various resources constraints.

It is assumed that there exists a host system to handle the high-level tasks in the robotic computation hierarchy. Since the I/O requirement of the IKS algorithm is very simple, the IKS chip can appear to the host system as a slave device. Figure 4-1 illustrates the "black box" specification of the IKS chip's interface. Note that some of the signals are for testing only and must be hardwired to proper values as indicated in the parentheses. I/O signals may be organized into three 16-bit words as shown in Figure 4-2. The host system obtains the IKS through read/write operations on these I/O ports following the procedure specified in Figure 4-3. The host is responsible for setting up the input transformation matrix and the arm configuration before asserting the signal *S* to start the IKS computation. In this way, the chip's interface to the host is not tied to a particular bus design and can be programmed easily. Incorporation of user-specified arm configuration in the IKS calculation is

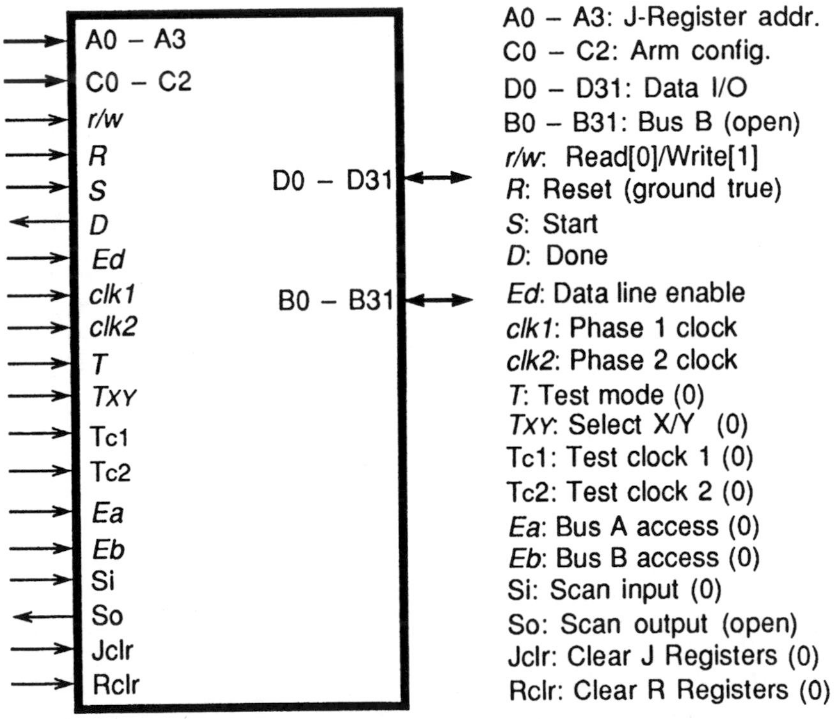
A0 – A3
C0 – C2
r/w
R
S
D
Ed
clk1
clk2
T
TXY
Tc1
Tc2
Ea
Eb
Si
So
Jclr
Rclr
D0 – D31
B0 – B31
A0 – A3: J-Register addr.
C0 – C2: Arm config.
D0 – D31: Data I/O
B0 – B31: Bus B (open)
r/w: Read[0]/Write[1]
R: Reset (ground true)
S: Start
D: Done
Ed: Data line enable
clk1: Phase 1 clock
clk2: Phase 2 clock
T: Test mode (0)
TXY: Select X/Y (0)
Tc1: Test clock 1 (0)
Tc2: Test clock 2 (0)
Ea: Bus A access (0)
Eb: Bus B access (0)
Si: Scan input (0)
So: Scan output (open)
Jclr: Clear J Registers (0)
Rclr: Clear R Registers (0)

Figure 4-1. The interface of the IKS chip.

D	R	S	r/w		Ed	C2–C0	A3–A0
D31–D16							
D15–D0							

Figure 4-2. Organization of the IKS chip's signals into three 16-bit words.

Step 1 : Set $R = 0$ for one cycle;

Step 2 : Set $R = 1$, $r/w = 1$, $Ed = 0$;

Step 3 : Repeat for $i = 0$ to 8 do   /* Input: write to J Registers */
  Write i to A_3–A_0 and the corresponding element to D_{31}–D_0
  according to the table below;

$A_3 A_2 A_1 A_0$	Jx	$D_{31} - D_0$
0 – 2	J0 – J2	*o* vector
3 – 5	J3 – J5	*a* vector
6 – 8	J6 – J8	*p* vector

  Set $Ed = 1$ for one cycle, then set $Ed = 0$;
 end repeat;

Step 4 : Set $r/w = 0$, and set $S = 1$ for one cycle, then wait for $D = 1$;

Step 5 : Repeat for $i = 9$ to 14 do   /* Output: read from J Registers */
  Write i to A_3–A_0 ;
  Set $Ed = 1$ and read θ_{i-8} from D_{31}–D_0;
  Set $Ed = 0$;
 end repeat.

Figure 4-3. Computing procedure using the IKS chip.

possible but not implemented here. Singularity cases are assumed to be handled by the host. In other words, the host system is responsible for the final interpretation of the results from the IKS chip. Since parameter values of the IKS algorithm vary only with different robot manipulators, it is more efficient to regard them as constants for repeated calculations and thus they are stored in ROMs. It is assumed that different parameter values can be written into the ROM in the final steps of the chip's fabrication.

Gate array technology is used since it is the preferred choice for designs that require a small production volume and rapid prototyping. This assumption is made to facilitate the estimation of area and delay time. The architecture, however, could also be implemented in standard cells or a full custom design. Basic cells and high-level macrofunctions, such as adders and multipliers, from commercial gate array libraries are used as the building block circuits of the IKS chip, and areas and delay times are estimated from databook values provided by the vendors. While the performance of the chip eventually depends on the physical design and the vendor implementation, it is nonetheless assumed that the architecture is independent of the technology in the sense that with a better technology (measured by, say, the achievable functional throughput rate), the architecture will always perform better.

4.1.2 Design Philosophy and Objective

A design philosophy is a set of values or stands regarding the approach to a design task. It is not necessary to judge that a particular philosophy be right or wrong, since it is largely a product of the interaction between the designer's objective and his/her environment (resources constraints and invested interests). On the other hand, however, knowledge of the design objective and philosophy is an integral part of the design decision making process, because they are the ultimate justifications of, or attacks on, a designer's decision. With this understanding, the design objective is first explained, followed by a discussion of the philosophical stands of this design effort.

One premise that leads to the decision of implementing the IKS algorithm in an ASIC chip is that future computation demands primarily come from novel applications, which require real-time response in an often unstructured environment. In this case, the latency rather than the throughput of the computation is the major concern. For example, if the robot is interrupted in the middle of a task execution along a predefined trajectory, real-time response requires that a new trajectory be calculated and executed immediately. Prolonged delay due to large computation latency may not be acceptable. Hence, the objective is to obtain an efficient chip architecture that will compute the IKS with minimal latency.

As an algorithm-specific processor, the IKS chip shares some limited common ground with general purpose processors. In this regard, it is important to learn from experiences in microprocessor designs while being aware of their limitations in algorithm-specific processor designs. In the arena of microprocessor design, from an architectural viewpoint and given the technology factor constant, there are basically three routes to achieve higher processor performance. The first is to increase the microprocessor throughput via parallel techniques such as pipelining or simply adding more functional units. The second is to reduce the number of instructions needed for a given algorithm. And the third is to speedup the execution of individual instructions by reducing the number of clock cycles per instruction and/or shortening the clock cycle itself. Ideally, all three approaches should be employed to achieve a higher performance, but in practice they are often incompatible as taking steps to enhance one aspect may produce adverse effects on the other. In fact, while the first approach is commonly accepted in today's microprocessors, the last two approaches may be viewed as the rationales behind the CISC (Complex

Instruction Set Computer) and RISC, respectively. This is because an ever-powerful instruction set is useful in reducing the number of instructions; the desire to speedup the clock rate, however, calls for simpler control decoding and thus a simpler instruction set.

After several years of debate, the dispute between the CISC and RISC has begun to subside. While the latest microprocessors such as the 88000 from Motorola and the 80960 from Intel have been developed based on the RISC concept, the current market is still dominated by CISCs [Col88a,b, Man88]. On the other hand, even "pure" RISC companies have incorporated CISC features in their latest product [Cus88a,b]. It seems that following the classic path of theory development, the phase of synthesis has begun to emerge now after the prolonged phase of thesis (CISC) and a relatively short phase of antithesis (RISC).

What can be learned from the debate between RISC and CISC? Central in the debate is the issue of instruction set design as the focus of the hardware resource allocation policy. A sound policy is founded on the principle that hardware resource allocation decisions must be based on a tradeoff analysis of the needs of the target application and the potential benefits of a specific decision. To a large extent, the argument between RISC and CISC is not about this principle, but about the perceived "needs" and "benefits" due to the different invested interests of their advocates. With this in mind, the design of the IKS chip should adhere to the above stated principle rather than siding purely with either RISC or CISC.

Another more profound reason for this stand is that despite their differences, both RISC and CISC are developed for general purpose applications. In this case, the analysis of the application's needs is usually based on a set of benchmark programs. Thus, the results of such analysis are valid only in a statistical sense. There is no guarantee, for example, that the instruction set is 100 percent useful for a particular algorithm even when a perfect compiler is assumed. In contrast, in designing algorithm-specific processors such as the IKS chip, the application algorithm is known prior to the design. Therefore, the application needs can be identified precisely. The benefits of a design decision can also be judged not by either the number of instructions (clock cycles) or the clock rate alone, but by both, i.e., the entire execution time. The implication is that if the instruction set is the focus of the hardware resources allocation, then the instruction set for the algorithm-specific processor not only can be "reduced" from complex instruction sets, but should be "derived" directly from the algorithm. This thinking leads to the philosophy of a "derived instruction set computer (DISC)" design and the development of an ASIC architecture design methodology that is not driven by a preconceived instruction set, reduced or not, but rather is aimed at deriving an architecture together with its behavioral representation — the instruction set — from the application algorithms.

4.2 An ASIC Architecture Design Methodology

The development of an ASIC architecture design methodology is an effort to cultivate a systematic approach to the discipline of algorithm-specific architecture design. In particular, it is aimed at developing a strategy that provides foci for hardware resources allocation decisions and enables these decisions to be made in a systematic manner. An overview of the methodology is given in the next section; its connection to the conceptual framework is also pointed out. Decision foci and guidelines for each design phase are then discussed in the subsequent sections.

4.2.1 Overview

In the ASIC design conceptual framework, it is suggested that recognition of design alternatives can be facilitated by the notion of the design space. Furthermore, the structures of the algorithm space and architecture space have been described. The algorithm space is decomposed into the three subspaces of operation, data structure, and data dependency. The architecture space is decomposed into the three subspaces of functional units, communication facilities, and control. The idea behind this characterization of the design space is that the matching of an architecture to the characteristics of the target algorithm can be decomposed likewise into the matching between the three pairs of corresponding subspaces as illustrated in Figure 4-4. With this picture, the design techniques of resource configuration and algorithm restructuring can be visualized as maneuverers that adjust the foci or projections in the architecture space and the algorithm space for a better match. An incremental design approach can be viewed as a process of making such adjustments continuously until a satisfactory solution is found.

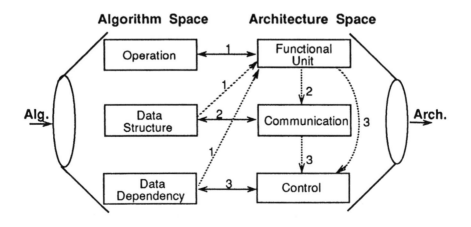

Figure 4-4. Matching the architecture to the task algorithm.

Figure 4-4 also illustrates one possible way to implement this algorithm-projection/architecture-configuration process. A designer may start from the algorithm characteristics of operation, data structure, and/or dependency to determine a particular architecture style and a set of functional units. With the functional units configured, the dataflow patterns are made explicit, and thus the communication facilities that support that flow pattern better can be determined. Finally, with both functional units and communication facilities taking shape, the control structure and mechanisms can be specified. This design path is presented as an ASIC architecture design methodology as illustrated in Figure 4-5. Essentially, the methodology divides the architecture configuration process into the three phases of functional unit configuration, communication configuration, and control configuration, with each phase having its own decision focus as denoted by the heavyweight frame. The paths where critical decisions are made in each phase are denoted by heavyweight arrow lines. It should be emphasized that even though the methodology presents these three phases in a linear order with limited interactions between these phases, iterations can be numerous in the actual process, depending on the designer's experience. In the next three sections, the decision focus and guidelines for each of the three hardware configuration phases are discussed in detail.

4.2.2 Phase 1: Functional Unit Configuration

Functional units are circuit modules of two broad categories: operation modules for arithmetic and logic operations, and storage modules for operands and results. As shown in Figure 4-5, the given application (task) algorithm to be implemented is the entry point of this design phase. To start, the designer must acquaint himself/herself with the algorithm and identify its characteristics in operations, data structure, and data dependency. From the identified characteristics, the designer may be able to select a basic architecture style and determine where the key design efforts should lie. An algorithm may have a very simple operation with regular data structure, or it may have very complex operations but no regularity at all in either its data structure or dependency. One may thus concentrate the design effort on storage modules in the former case, and on operation modules or the control structure in the latter case. On the surface, decisions in this phase may not appear to be particularly hard to make since most of the necessary building block modules are probably available from design libraries. But the decisions are crucial in determining the effectiveness of the final design as they affect decisions made in the two subsequent phases. Once these decisions are made, the task algorithm can be translated into a pseudocode representation for further work in subsequent phases.

4.2.2.1 Decision Focus: The Functional Unit Profile

The functional unit profile contains the essential information about how operations elementary to the task level are to be realized. It thus serves to establish a link between the physical hardware entities and the symbolic objects through which an algorithm is expressed. Information contained in the

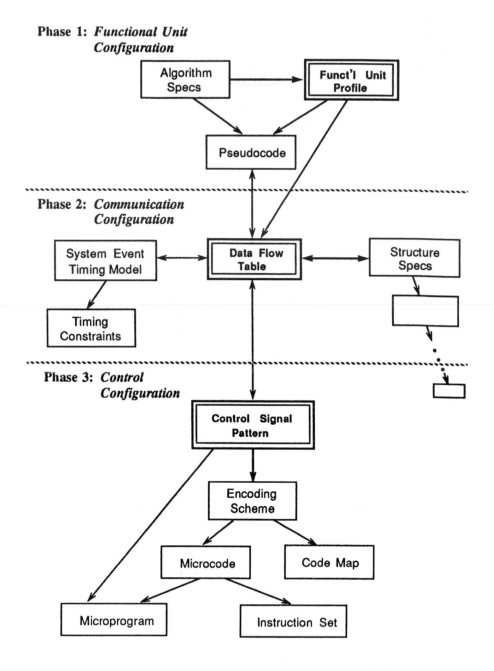

Figure 4-5. The ASIC architecture design methodology.

functional unit profile is divided into three categories. The first category describes the data formats used by various circuit modules. This may be viewed as a translation of the software concept of data typing into hardware. The second category describes circuit modules to be used. This part can be further divided into operation modules and storage modules. The third category describes certain "macros" that specify how the higher level operations (patterns) are translated into groups of operations. For example, in the functional unit profile for the IKS chip, two macros of *mac* and *cordic* are present to describe how certain operation (patterns) are translated to the multiplier-accumulator operation and the cordic operations, respectively.

The functional unit profile forces the designer to focus his/her attention on the decisions about circuit modules. It also creates a symbolic name table for hardware objects to be further manipulated in later design phases. CAD tools can be developed to automate routines such as bookkeeping functions associated with the profile information management and the translation of the task algorithm into pseudocode.

4.2.2.2 Alternatives and Guidelines

Decisions on the choice of functional units have two aspects: the architectural (global) aspect and the algorithmic (local) aspect. The term algorithm has two implied meanings: task algorithm refers to the high-level algorithm to be implemented (e.g., the IKS algorithm in this work); hardware algorithm refers to the algorithm that a circuit module implements to realize operations elementary to the task level (e.g., trigonometric functions, square rooting, etc.). The algorithmic aspect here refers to the effects of choosing a particular hardware algorithm.

The experience of this work indicates that in making decisions on functional units, the architectural aspect, which has a global effect, should be considered first. And, it should be considered from the characteristics of the task algorithm. The key issue is the choice of an architecture style that looks more promising given the amounts of parallelism and regularity exhibited in the task algorithm. Metrics for measuring the effectiveness of matching an architecture style to a task algorithm have been discussed at length in Section 2.2.4.1. It provides the first basic guideline to follow in making functional unit decisions.

More local algorithmic aspects include issues of what and how primitive operations are to be implemented in hardware, as well as the effects of the I/O requirements, data structure, and data formats. Some basic guidelines can be stated:

- Keep the I/O and data formats simple;
- Keep the number of data formats small;
- Choose functional units that can be shared by other hardware algorithms;

- Use well proven designs;
- Assign a higher resource priority to primitives that are used more frequently.

Because of the local nature of the issues they address, these guidelines should be applied in the context of the particular design situation. For example, a less frequently used primitive may have greater impact on the performance, or idle resources may be available for performance improvement even though a primitive may have lower resource priority. The emphasis may shift from case to case. Therefore, the designer must exercise common sense judgements in applying these guidelines.

4.2.3 Phase 2: Communication Configuration

Communication facilities include interconnection schemes and the data transfer mechanisms such as synchronization and timing. Interconnects can be divided into two classes of busses and direct links. Busses are characterized by multiple sources and destinations. In processor designs, internal busses are usually based on simple synchronous schemes and the routes are directly set up by the control signals. Also, precharge mechanisms are often employed to improve the data transfer performance. This is because the length of the wiring and the number of connections usually make capacitive loading large. By precharging the bus before a data transfer, the individual driver cells need only to discharge the bus. Direct links, on the other hand, are characterized by a single source and destination pair. They typically are not precharged and their performance is thus inevitably affected by the physical placement of the connecting modules. In addition to interconnections, communication also depends on the timing of both the data and control signals.

The communication needs of an application algorithm depend on the dataflow patterns for a given set of functional units. To make the dataflow patterns explicit, the pseudocodes specified from the previous design phase are translated into a dataflow table. The timing model for system events is first developed. From the timing model and the desired dataflow pattern, an initial interconnection scheme may emerge. Aspects of the interconnections, operations, and timing may depend on the feasibility of the implementation. Thus, hierarchical decompositions and detailed designs of circuit modules may proceed in parallel. The detailed circuit design, the timing model, the interconnection scheme, and the dataflow patterns are intensely interwoven. As the design proceeds further into more detail, the designer may find it necessary to switch from one of these facets to another, or even from one design phase to another. This "context switching" is facilitated by the representation of the design in a dataflow table, which essentially packs all the logical aspects of various design facets into a two-dimensional structure and thus enables the designer to maintain multiple views without changing the representation. As shown in Figure 4-5, the dataflow table plays the key role as the central representation of the logic design; consistencies among specifications in other design facets are maintained by the consistencies between the dataflow table and each facet. Working with this central representation, the designer can manipulate the dataflow and refine the interconnection scheme, timing, and detailed circuit designs simultaneously.

4.2.3.1 Decision Focus: The Dataflow Table

Perhaps the most versatile software tool created by the recent microcomputer revolution is the spreadsheet program. These programs turn the task of managing multiple dimensional information into simple editing. The dataflow table is created to take advantage of such a tool by displaying information relevant to the dataflow of an algorithm in a two dimensional structure. Specifically, the time dimension of the flow naturally coincides with the horizontal direction and has the basic unit of instruction cycle. The cycle can be further partitioned into phases depending on the timing model. In the vertical direction, there are three kinds of information present. The first contains information that serves as linkage between the current phase and other design phases. This includes an index of the task to which the current cycle belongs, the indices of instruction cycle and clock cycle. The second kind contains information about the execution details such as where to get and how to transfer the operands and what operations to perform. The third one contains other auxiliary information such as comments regarding some of the particular details of that cycle. Static design information, such as the existence of an operation module or an interconnection, is embodied in the structure of the table and is thus suppressed. This makes the dynamics of the dataflow, specified by the table entries, more visible. Frequent dataflow patterns are thus more easily recognized. Interconnection schemes that facilitate those dataflow patterns can be derived. Further finetuning of the interconnection scheme involves local manipulations of the dataflow. This can be accomplished easily by using the built-in table editing functions of the spreadsheet program.

4.2.3.2 Alternatives and Guidelines

For general processor design the goal of communication configuration is to provide a data transfer bandwidth large enough to support full utilization of the functional units, thus providing maximum throughput. Such an objective is reasonable in the general case only because the exact dataflow is not known during the processor's design. In the algorithm-specific processor design environment, however, this underlying fact no longer exists. Not only is the exact algorithm known, but also the exact dataflow can be manipulated from both the algorithm and architecture sides. Since communication facilities involve the interconnection scheme, timing, functional circuit designs, and the supporting control mechanisms, idle resources in one facet may be traded for improvements in the other. Instead of trying to obtain a maximum throughput, a more productive goal is thus to negotiate a compromise that balances these various concerns. This can be done by manipulating the dataflow through the dataflow table. In the remainder of this section, some options available for this maneuvering are discussed.

Restructure of the Algorithm

This includes techniques of changing the assignment of registers, changing the operand routing, scheduling a different order of subtask executions, using idle functional units for partial computations, etc. In some situations, the result of the manipulation may introduce extra instruction cycles in the execution. If the resultant cycle(s) does not increase the control signal patterns or exceed the program space available, then it is generally acceptable.

Hardware Implementation

Many operations such as transfer of a constant, truncation or shifting of an operand, and branching can be implemented by hardwiring techniques. In most cases, the added area and control complexity are usually not a problem, but if the hardwired operation is on a critical path that determines the clock rate, then it may affect the entire computation latency.

Addition of Direct Links

Each addition of a direct link will generally increase the total possible bandwidth of the system. The decision is obviously determined by the added link's utilization as well as the potential enhancement of dataflow. The designer should also be aware of the implications on the physical placements of circuit modules.

4.2.4 Phase 3: Control Configuration

As the communication configuration phase concludes, the processor's execution of the task algorithm must now be translated into an exact dataflow stream as specified in the dataflow table. This detailed behavior of the processor is to be realized by various system events whose occurrence in turn depends on the control vector, i.e., the collection of the control signals. An instruction is then a behavioral interpretation of the control vector value. The collection of all valid control vector values forms the processor's instruction set.

New opportunities have been created in the design of instruction sets for algorithm-specific processors. Taking advantage of these opportunities relies on recognizing the control signal patterns since instructions are no more than the behavioral interpretation of the control signals. While the control signals are implicitly determined by the dataflow, their patterns are obscured by other information in the dataflow table. To make these patterns more explicit, the control signals are compiled and sorted from the dataflow table to form the control signal pattern profile. Based on an analysis of the alternatives in implementing the patterns, an encoding scheme can be determined, possibly accompanied by some further dataflow manipulations. The control signals are then specified as microcodes and recast as the processor's instruction set. The task algorithm is then transformed into a microprogram compiled from the dataflow table using the processor's derived instruction set.

4.2.4.1 Decision Focus: The Control Signal Pattern Profile

Recognition of any pattern is facilitated by the categorization of the subject phenomenon. The control signals of the processor can be categorized in a number of ways. For example, they can be categorized in terms of their timing as phase one or phase two, or according to the modules to which they belong. In this work, the control signals are categorized in terms of their clock phase so that they are roughly divided into the two categories of data transfer routing setup and operation control. A control signal pattern, consisting of values in these two categories, is essentially equivalent to a control signal vector without address specifications. To assist modifications, each pattern is associated with a link list that allows a pattern to be traced back to the corresponding instruction cycles in the dataflow table. Furthermore, the values of the two categories in the pattern are specified by a mixture of symbolic names and the actual signal values so that the consistency can be checked by inspection. Again using the spreadsheet software, the patterns are then sorted and regrouped to make the pattern explicit.

4.2.4.2 Alternatives and Guidelines

For algorithm-specific processor designs, the control vectors for the dataflow are likely to require relatively small program space. A simple on-chip control store approach may be possible. This approach is most advantageous since off-chip propagation delays are eliminated from control signals and more I/O pins can be used for data transfer to speed up the testing, etc. This approach also favors the separation of the control and data storage, the so-called Harvard architecture, since the word size requirements of the two are likely to be different.

Areas for the on-chip control store can be reduced by encoding the vector values. The encoding can be carried out in multiple levels, but each level will introduce a decoding delay into the control signal propagation. The delays due to the decoding, however, may or may not affect the performance, depending on whether the control signal decoding becomes the critical delay path. If the circuit design is decomposed hierarchically along the functional line, then it is generally desirable to have the control signal relevant to the circuit module decoded locally since the delay involved is usually tolerable and routing space can be saved.

The ASIC architecture design methodology prescribes a two-level structure for implementing the execution control of the task algorithm: the control signals are encoded as microcodes, and a particular sequence of the microcodes forms the microprogram. The encoding space, formed by the unique control vectors that can be represented by the code, has a size of 2^n, where n is the number of bits used by the code. Note that each increment of the code size induces a power of 2 jump in the encoding space. (This phenomenon is also true for the storage space, including the program space). The encoding efficiency can be measured by the ratio between the number of defined microcodes and the encoding space size. The number of control signal patterns in the profile produced from the initial dataflow may not exactly match the encoding space size. This is the point where the CISC or RISC comes into the play. If the number of patterns are slightly below some quantum value, then the CISC approach may be taken to manipulate the dataflow so that the total cycles are reduced. On the other hand, if the number of patterns just exceeds some power of 2 value, then the RISC approach may be used to get rid of some infrequently used patterns, possibly at the expense of the total clock cycles. Alternatives in encoding schemes may be aimed at minimizing space or time, and the choice depends on the overall need of the design. Since the encoding scheme is developed from the control signal pattern profile, it treats the patterns as if they are unrelated. After the microprogram is compiled, additional opportunities for minimizing the area by exploiting some patterns or relationships in a microcode subsequence may be possible.

4.3 The IKS Chip Architecture Design

The IKS algorithm to be implemented is specified in Appendix A. The architecture design of the IKS chip is presented as the result of the execution of the ASIC architecture design methodology described in the last section. That is, the design process is viewed as a series of hardware resource allocation decisions that lead to the final design. Alternatives and the evaluation of them are emphasized throughout the decision making process.

4.3.1 Design Decisions on Functional Units

Decisions on functional units for the IKS chip are made as follows. First, the characteristics of the IKS algorithm are identified. Architectural and algorithmic alternatives are then explored and evaluated. The evaluation leads to the development of an architectural idea, which becomes the root of the circuit hierarchy. Following the top-down approach, operation details of each module are refined and specified in the functional unit profile.

4.3.1.1 The Characteristics of the IKS Algorithm

The IKS algorithm is analyzed in terms of operation, data structure, and data dependency. For operation characteristics, the IKS algorithm has very simple I/O requirements (9 input data and 6 output data) provided all computations can be accomplished on chip. On the other hand, the computation requires a rich set of arithmetic operations including trigonometric functions and square rooting. No explicit logical operation is required. Table 4-1 lists the operation counts in various categories. In the first column, the strict counts of multiplications and addition/subtractions are shown. In the second column, multiplications and additions that are of the form ab+cd are counted separately. This form of computations can be efficiently executed in the well known multiplier-accumulator (MAC) structure. However, the statistics clearly indicate that neither the trig functions nor the MAC operations alone dominate the entire IKS calculation.

Table 4-1. Operation Types and Counts in the IKS Computation.

Type	Strict Counts	With MAC
add/subtract	25	7
multiply	35	1
ab+cd	–	17
sine	5	5
cosine	5	5
arctangent	7	7
square root	2	2

Data structure can be identified from a macro or micro view. From a macro view, while the input data are vectors, the algorithm itself is highly irregular except that most of the multiplication and additions are in the form of ab+cd. From a micro view, there are four data types: orientation elements, position elements, angles, and the squares of position elements. The last type is transparent to the user. The exact data formats of these data types are defined in the functional unit profile.

The data dependency is often presented as a directed graph, with nodes and arcs representing the operation tasks and dependency between the tasks, respectively. Lee and Chang have decomposed the entire IKS computation into subtasks that are individually executable in a single cordic step. Figure 4-6 shows the directed graph of the IKS algorithm in this formulation [LeCh87]. Inspection of Figure 4-6 reveals that the IKS algorithm has a strong sequential data dependency with a scarcity of parallelism. This is not surprising as it is only a reflection of the nature of the joint-link structure in the robotic arm.

The combination of simple I/O requirements, strong data dependency, and a lack of inherent parallelism and structural regularity in the IKS algorithm leaves little room for performance improvement through massive parallel techniques. Since computation of trig functions and square roots is usually slower than simple add/subtract operations by an order of magnitude, and since these operations constitute a substantial portion of the total computation needs, the effectiveness of the architecture will be determined to a large degree by how the trig functions and square rooting are implemented. Through this initial analysis, it is recognized that the decisions on operation module designs are the most critical ones.

4.3.1.2 Basic Architectural and Algorithmic Alternatives

In this section, unless otherwise specified, the term algorithm refers only to hardware algorithms. The strategy is to first consider the architectural alternatives based on the analysis of the task algorithm in the previous section. The understanding of architecture alternatives is then used to guide the decisions on algorithmic alternatives.

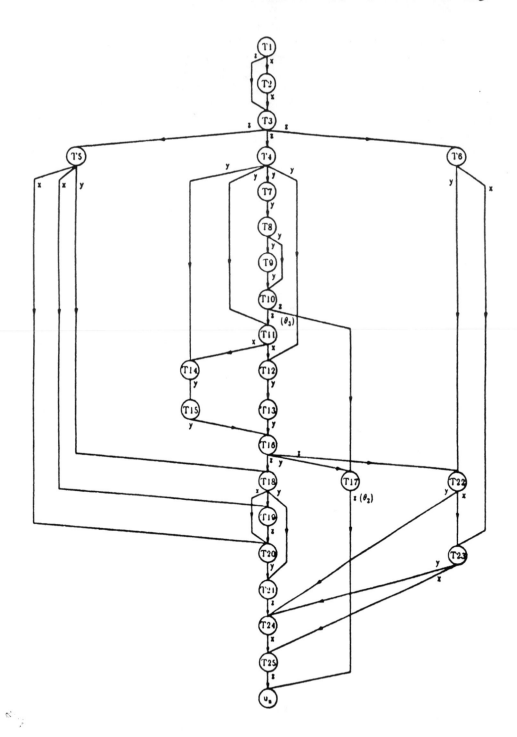

Figure 4-6. The task graph of the IKS algorithm [LeCh87].

One architecture alternative is the choice between using single processors or multiprocessors. The latter can be further divided into two subclasses of heterogeneous and homogeneous. The lack of inherent parallelism in the IKS algorithm means that addition of more processors will not speedup the computation very much. Moreover, the architecture will become less effective as the number of processors increases. Therefore, the only effective alternative for the multiprocessor approach is to use a coprocessor to speedup the trig functions and square rooting. But this approach will destroy the I/O simplicity and is potentially very costly unless the coprocessor itself can be incorporated on the same chip. As a result, the multiprocessor approach is quickly rejected.

If the single processor approach is adopted, then the processor must be able to handle trig functions and square root calculations efficiently. This stresses the importance of finding an efficient algorithm to implement all the needed arithmetic operations. One particularly attractive alternative is the cordic algorithm [Vol59, Wal71, HaTu80, Sco85, CaLu87]. It has been demonstrated that the IKS can be computed by using only cordic processors [LeCh87]. However, a closer look at the underlying cordic algorithm reveals that each cordic cycle consists of a number of basic adder cycles depending on the word size. As a result, a raw cordic processor is inefficient in computing the additions and multiplications. Since a substantial portion of the IKS computation requires multiplication and addition, using a cordic processor alone implies a heavy penalty on those simple arithmetic operations.

Another candidate for the single processor approach is to use an MAC structure. A dedicated multiplier can speedup multiplications, and the MAC structure is especially advantageous for operations of the form $ab + cd$. Also, these operation modules are readily available in the libraries from ASIC vendors, and since they are usually optimized and reliable, this choice has an implementational advantage. With the MAC, however, the computations of the trig functions and square roots are quite involved. Various algorithms that combine hardware and software techniques to compute these functions efficiently have been proposed [Pen81, Tay81, Far81, HwWX87, BPTP87]. These techniques typically involve either iterations to save hardware, or massive tables to obtain an answer through lookup and interpolation. Because of the many different types of functions that need to be computed, the alternative of table lookup technique will require too large an area and is not practical for a single chip implementation.

The most troublesome problem in using the MAC structure is the computation of the arctan function. The common method to compute the arctan is based on power series expansion. This immediately introduces another unwanted operation, division. Moreover, to save operations, the power series is usually computed recursively under the form of $x_{k+1} = a_k + b_k{}^*x_k$. With this formulation, the operands for the next step of computation depend on the results of the current step. This creates a dilemma for the execution in a pipeline design. On the one hand, if one increases the clock rate through partitioning the multiplier into stages, with the delay of each stage comparable to the adder, empty cycles have to be inserted into the iteration steps. This results in more clock cycles being needed for the entire computation. On the other hand, if the clock rate is made equal to that of the multiplier, the simple arithmetic operations are penalized. But the most severe disadvantage of the MAC approach is that unlike cordic processors, algorithms that implement individual trig functions or square rooting are difficult to combine. As a result, control resources needed for these algorithms may not be shared and design effort for implementing each of these functions becomes isolated.

Three alternatives of coprocessor, cordic processor, and an MAC structure have been discussed. Each of these alternatives has its own advantage, but none is satisfactory. This suggests that a synthesis of the three alternatives may provide a more cost-effective solution if a way can be found to exploit the advantages while avoiding the unfavorable aspects of each alternative. This insight leads to the idea of an architecture featuring a cordic core embedded into an MAC structure as illustrated in Figure 4-7. This architecture is named MACC for Multiplier-Accumulator with a Cordic Core. The cordic core and the MAC share an adder as indicated by the overlapped area. The cordic core can be viewed as an on-chip coprocessor or, more aptly, a hardware subroutine that is responsible for all trig and square root functions. The MACC can also be viewed as a cordic machine incorporating an on-chip multiplier coprocessor. A quick estimate based on operation counts indicates that the MACC may be able to achieve a speedup factor of 1.5 over the single cordic processor implementation. Before the functional unit profile of this architecture is described, a piece of information about the factors that will affect the accuracy of the computation is needed.

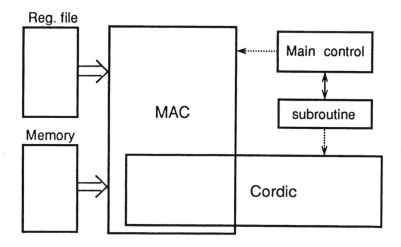

Figure 4-7. The architecture concept of MACC.

4.3.1.3 Simulation Results of Fixed-Point Calculation

Surveys have shown that the positional resolution requirements of various robotic applications range from 0.1–10.0 mm for a working range of 1–2 m [Har82]. The resolution requirement can be satisfied by a 14-bit value. Repeatability requirements of 0.01-1 mm are typical [Jar84]. These requirements suggest that the IKS computation can be carried out in fixed-point calculations, resulting in tremendous saving in hardware resources.

To obtain a quantitative assessment, a simulation program written in C has been developed to investigate the relationship between the computation error and the word size. The simulation flow of the program's main loop is illustrated in Figure 4-8. The program first randomly generates a joint angle

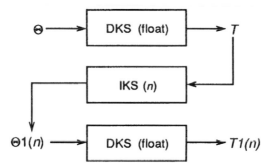

Figure 4-8. The flow of the fixed-point simulation program.

vector (Θ) and calculates, using floating point numbers, the corresponding position and orientation vectors in Cartesian coordinates (T), which become the input to the IKS program module. The IKS program then computes the joint angles ($\Theta1(n)$) using fixed-point calculations. Both the input and all the intermediate calculations are truncated to a word size of n bits where n varies from 20 to 24 bits. (In an earlier investigation on direct kinematics calculations, it was found, somewhat surprisingly, that the simple truncation scheme performs consistently better than two other leading rounding schemes. Therefore, simple truncation is used in the investigation on IKS.) The new position and orientation vectors ($T1(n)$) corresponding to the joint angle vector obtained from the IKS program are also computed. Errors between Θ and $\Theta1(n)$, and between T and $T1(n)$, for n = 20 to 24, are computed. A counting criterion ε is defined for the error of position elements for each word size. The maximum error (within the counting criterion) in each category is recorded and the error of each element is counted in terms of the achieved accuracy (in bits). Errors greater than the criterion are also counted separately but are not collected in the maximum error record. The statistics of $\Theta1$ (qi's) and $T1$ (orientation and position elements) from the simulation on 10,000 joint angle input vectors are shown in Table 4-2.

Table 4-2. Accuracy in Fixed-Point IKS Calculation.

Word Size	Max. θ error (rad)	Max. ori. error	Max. pos. error (mm)	Criterion ε (mm)	Count (error>ε)
20	0.0400	0.0017	1.4644	1.6	12
21	0.0259	0.0009	0.7901	0.8	9
22	0.0080	0.0005	0.3863	0.4	9
23	0.0028	0.0002	0.1996	0.2	9
24	0.0012	0.0001	0.0946	0.1	10

The statistics show that if the IKS is computed using fixed-point numbers, the error due to truncation may not be acceptable for a word size below 20 bits. It also shows that as the word size increases by one bit, the accuracy distributions retain the same shape but generally shift one bit towards a higher accuracy as expected [Leu89a]. The maximum values recorded in the intermediate steps indicate that overflow may occur. But the overflow condition can be eliminated by simply increasing the word size by one bit. Further simulations of the cordic operation indicate that an extra bit is needed to prevent overflow. Thus, if the position resolution requirement is 0.1 mm, and if an error rate of 0.1% is acceptable, then a word size of 26 bits is adequate.

4.3.1.4 MACC Functional Unit Profile

The analysis presented in the previous three sections indicates that the MACC with fixed-point computation promises a cost-effective architecture for computing the IKS. This architecture idea is transformed into more concrete specifications through the development of the functional unit profile. The MACC functional unit profile consists of three parts: data formats, circuit modules, and macros.

Data Format

It has been indicated that an additional $\log_2(n)$ bits are required internally to achieve an n-bit accuracy in the cordic [Wal71]. For the desired accuracy of 26 bits, 5 extra bits are needed. Since 32-bit adders are readily available in most ASIC libraries, the basic word size is fixed at this value. There are four data formats used in the IKS calculation, but only the first three are visible to the user. The four data formats are defined in Figure 4-9.

The value of an angle used in the IKS calculation is a number representing a fraction of a turn. In this representation, the first two bits together indicate the quadrant in which the angle falls. A value of 0.5 thus indicates an angle of 180 degrees. The cordic core will operate on quadrants I and IV only; angle values in the II and III quadrants must be mapped into the I and IV quadrants during a prepossessing step. Another postprocessing step will map the results of the

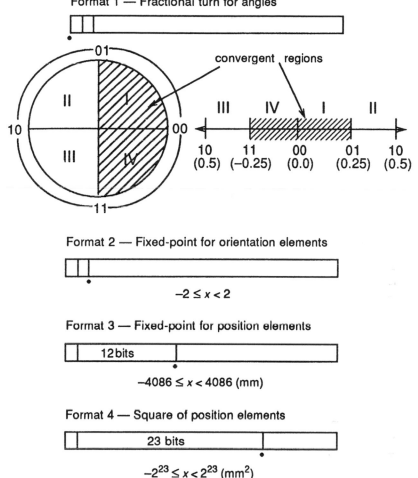

Figure 4-9. The four data formats used in the IKS calculation.

cordic operation back to the correct quadrants. The advantage of this representation is that the preprocessing and postprocessing of the angle values can be implemented in hardware with a small area overhead and the processing can be carried out while the data is "on the fly".

Circuit Modules for Operation

The MACC is composed of an MAC and a cordic core. Three adders are needed in the cordic algorithm. Since the IKS algorithm has strong sequential data dependency, concurrent execution of MAC and cordic is not particularly useful. Therefore, one of the adders from the cordic core can be shared with the MAC to save area. The basic arrangement of the three adders is shown in Figure 4-10.

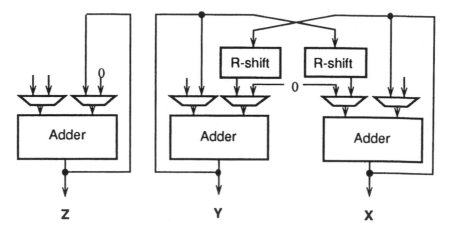

Figure 4-10. Block diagram of cordic implementation.

An inspection of the figure suggests that the Z adder should be designated as the shared adder since the other two adders are more logically related. Moreover, the X and Y adders are active only in executing the cordic algorithm; accordingly, they are treated as a single module. Since only three of the full cordic functions are required by the IKS algorithm, unused functions (such as division and hyperbolic functions) and their supporting resources are stripped from the full cordic implementation. A mode variable is used to specify the three functions performed by the cordic core. The algorithm implemented by the cordic core is specified in Figure 4-11. To smooth the transitions between cordic and non-cordic operations, the preprocessing and postprocessing operations are handled by the main control. The core thus needs only to fetch the constants and handle the shifting and add/subtract controls during the iterations.

A 24-by-24-bit two's complement multiplier is used to generate a 32-bit product. When used as input to the multiplier, the least significant 8 bits of a datum are truncated. To maintain the proper data format, the most significant two bits and the least significant 14 bits of the result are discarded.

Circuit Modules for Storage

The numbers of constants needed by the cordic operations and parameters of the robot arm are 26 and 6, respectively. They can fit exactly in a 32-word ROM. To simplify their addressing, the cordic constants are stored from the ROM address 0 so that the iteration cycle index can be used as the fetching address.

The J register file contains 16 words and is accessible from the outside. The upper 9 words hold the input transformation matrix. The next 6 words store the joint angle values. The last word is reserved for future use.

The R register file is used as a scratch pad in the IKS computation and is transparent to the user. After optimizing the register usage, only 8 words are needed for the R register file. The separation of the R register file from the J

mode 1 (sqrt1)	mode 2 (sin & cos)	mode 3 (sqrt2, atan)

Step 1: Initialization and Range Reduction

mode 1 (sqrt1)	mode 2 (sin & cos)	mode 3 (sqrt2, atan)
$x := x_0, y := y_0$;	$x := k, y := 0, z := \theta$;	$x := x_0, y := y_0, z := 0$;
if $x - y < 0$ then	if $0.25 < z < 0.75$ then	if $x < 0$ then
set flag;	$z := z + 0.5$;	$x := -x, y := -y$;
end if;	set flag;	set flag;
	end if;	end if;

Step 2: Iteration for i = m to $n-1$ where m = 1 for mode 1, and

m = 0 otherwise

Ø1:   fetch α_i;

    compute y_i', x_i' : $y_i' := y_i \cdot 2^{-i}$, $x_i' := x_i \cdot 2^{-i}$;

    compute d_i: if $m = 2$ then

                $d_i := \text{sign}(z)$;

          else

                $d_i := -\text{sign}(y)$;

          end if;

Ø2:   compute in parallel: $x_{i+1} := x_i - s \cdot d_i \cdot x_i'$; where $s = -1$

                       $y_{i+1} := y_i + d_i \cdot x_i'$;   when mode = 1

                       $z_{i+1} := z_i - d_i \cdot \alpha_i$;

Step 3: Post-Processing

mode 1 (sqrt1)	mode 2 (sin & cos)	mode 3 (sqrt2, atan)
if flag set then	if flag set then	if flag set then
$x := 0$;	$x := -x; y := -y$;	$z := z + 0.5$;
else	end if;	end if;
$x := x \times k_h$;		$x := x \times k$;
end if;		
Output:	Output: $x = \cos\theta$	Output:
$x = \text{sqrt}(x_0^2 - y_0^2)$	$y = \sin\theta$	$x = \text{sqrt}(x_0^2 + y_0^2)$
		$z = \theta = \text{atan}(y_0 / x_0)$

Figure 4-11. The algorithmic description for the three modes
of the cordic core.

register file is primarily based on performance considerations. This is because two smaller registers would have simpler decoding and less loading than if they are combined. This results in a faster response. Moreover, the role of the J register (I/O) is different from that of the R register. The separation of the two allows a more focused performance enhancement objective for each circuit module and eases the design effort. Another reason is that since eight operands are potentially needed for keeping all the operation modules busy, separate registers provide additional transfer facilities. The price paid is two more address bits for the control word. This is considered justifiable in view of the potential performance gains.

Macro

Two macros of mac and cordic are defined for translating groups of operations in the algorithm into special hardware primitives. Their definitions together with the data formats and the circuit modules are summarized in Table 4-3.

4.3.1.5 The IKS Pseudocode

As shown in Figure 3-3, at the highest level of the VLSI design hierarchy, the behavior of a processor IC is represented by its instruction set. The pseudocode is an intermediate representation between the task algorithm usually specified in a high-level language and the chip's yet to be specified instruction set. With the MACC functional unit profile specified, the IKS algorithm is translated into a pseudocode program. The IKS pseudocode program is listed in Appendix B. To facilitate the subsequent design, test, and debug processes, the MAC and cordic operations are treated as independent tasks and labelled. Their corresponding roles in the IKS algorithm are also indicated by the equation numbers. Another C program is developed to simulate the execution of this pseudocode program. The correctness of the pseudocode program is verified and the computation details of each task are recorded, which become the test data for the later VHDL simulations.

The completion of the pseudocode specification of the task algorithm signifies the end of the first design phase, and the attention now shifts from finding an appropriate set of functional units to interconnecting them.

4.3.2 Design Decisions on Communication Facilities

Decisions on communication resources largely depend on understanding the dataflow patterns. To make these patterns explicit, the pseudocode program from the last design phase is translated into a dataflow table. Heuristic interconnection schemes are developed and improved upon incrementally. A timing model is simultaneously constructed and refined. Detailed designs and decompositions of functional circuit modules also proceed in parallel. The consistency among the interconnection scheme, timing, and data dependency of the algorithm is maintained through the dataflow table. The dataflow manipulation, the interconnection configuration, the timing specifications, and the detailed design of the functional modules are highly interrelated. Because of

Table 4-3. Summary of the MACC Functional Unit Profile.

Category	Specification			
Data format	Name/Notation	Representation	Range	Resolution
	angle: x_a	32-bit fractional turn	$0 \leq$ x_a $< 2\pi$(rad)	$\pi/2^{-31}$(rad)
	orientation: x_o	32-bit 2's complement	$-2 \leq$ x_o < 2	2^{-30}
	position: x_p	32-bit 2's complement	$-4086 \leq$ x_p < 4086(mm)	2^{-19}(mm)
	pos-square: x_s	32-bit 2's complement	$2^{-23} \leq$ x_s $< 2^{23}$(mm^2)	2^{-8}(mm^2)
Circuit module	Name	Operation	Input	Output
	MPY	P := m1*m2	m1, m2: the most significant 24 bits of either x_o or x_p, m1, m2: both in x_o format m1, m2: both in x_p format m1, m2: one in x_o, one in x_p	P: 32 bits [45:13] of the 48-bit full prod. [47:0] $\Rightarrow$ P in x_o $\Rightarrow$ P in x_s $\Rightarrow$ P in x_p
	Z adder	Z := z1 $\pm$ z2	z1, z2: 32-bit, same format	Z: same format as z's
	X adder	X := x1 $\pm$ x2	x1, x2: 32-bit, same format	X: same format as x's
	Y adder	Y := y1 $\pm$ y2	y1, y2: 32-bit, same format	Y: same format as y's
	M	Read-only memory, 32-word, 32-bit/word		
	J Register	I/O register, 16-word, 32-bit/word		
	R Register	internal, 8-word, 32-bit/word		
Macro	Name	Operation	Input	Output
	cordic (1,x,y,–)	r := sqrt(a^2–b^2)	x = a, y = b; x, y in same format	X = r in same format as x and y
	cordic (2,x,y,z)	s := sinθ, c := cosθ	z = θ in x_a format	X = c, Y = s, both in x_o
	cordic (3,x,y,z)	θ := atan(b,a) r := sqrt(a^2+b^2)	x = a, y = b; x, y in same format	Z = θ in x_a format X = r in same format as x and y
	mac (a,b,$\pm$,c,d)	Z := ab $\pm$ cd	a,b,c,d-24 bit a,b,c,d,Z: conforming to the formats specified in MPY and Z adder	Z: 32-bit

its pivotal role in the entire design process, the dataflow table and its manipulation are described first. The timing model and detailed timing specifications are presented, followed by a discussion of the decisions regarding the interconnection scheme of the IKS chip.

4.3.2.1 Editing the Dataflow Table

The IKS dataflow table is created and manipulated using a commercial spreadsheet program. Figure 4-12 shows a segment of the IKS dataflow table. For each cycle (column), the upper part identifies the corresponding task, instruction cycle, and clock cycle. The resource part has two buses, one multiplier, and three adders. The results of the multipliers and the three adders are identified by P, and X, Y, Z, respectively. Because of the two-phase clocking scheme (see next section), the columnar space of each table entry is partitioned into the left half and the right half corresponding to the two phases. Operand specifications are assigned in the left half (first phase) and the function types are specified in the right half. Entries for the busses are specified by a "source : destination" pair.

The initial dataflow of the IKS execution is generated by translating the pseudocode program into the table entries in two steps. In the first step, the operations of each task are carried out by assigning the name of the operands (such as register words identified by Jx or Rx, or the results of other operation modules in the previous cycle) to the appropriate operation modules in each

Task	T-27					T-28		
I-cycle	I-85	I-86	I-87	I-88	I-89	I-90	I-91	I-92
Clock	C-290	C-291	C-292	C-293	C-294	C-295	C-296	C-320
BusA	J2:m1	R4:m1		R1:y2		R5 : y2	Z : z1	
BusB	R3:m2	R2:m2					Mi : z2	
m1	J2	R4						
m2	R3	R2						
z1	P	P	P	P		0	Z	0
z2	0	Z	0	Z		0	Mi	ang(Z,f)
Zc	Z: R1 (+)		(−)Z: R6 (+)	(+)	Z : R4	(+)	Az	(+)
x1						0	X	
x2						Y	Y	
Xc						(f)	Ax	
y1				0	Y	0	Y	
y2				R1	0	R5	X	
Yc				(+)	(+)	(f)	Ay	
comment				setf=1			cordic,3	

Figure 4-12. A segment of the IKS dataflow table.

cycle without concern to how these operands are transferred. In the second step, from the observations on some basic dataflow patterns, an initial interconnection scheme is adopted and operands are heuristically assigned to the interconnections available. (The development of interconnection scheme is explained in detail in Section 4.3.2.3.) In this initial dataflow table, some operand transfers may be left unassigned due to a lack of data transfer resources.

The initial dataflow table is then examined to observe patterns of operand transfers and levels of resource utilization. To specify the remaining operand transfers, the table is edited. The dataflow manipulation essentially negotiates a compromise among various means of utilizing available idle resources, locally restructuring the task algorithm, addition of interconnections or cycles, and/or particular hardware implementations. This follows the decision guidelines discussed in Section 4.2.3.2. This process is repeated concurrently with the timing specification and the detailed designs of the module circuits until the interconnection scheme implied by the dataflow table satisfies the data transfer needs with reasonable utilization.

4.3.2.2 Timing Model of System Events

Because of testability considerations, the IKS chip follows the level sensitive design discipline with the synchronization of system events based on a two-phase non-overlapped clocking scheme. Level-sensitive dynamic latches, controlled by either clock phase, are used to partition a circuit module such that only combinational logic appears between latches. Level sensitive refers to the circuit behavior that the output signal of a latch responds to the value change (logic level) of the data input, after some delay time, as long as the control signal is enabled. In other words, the steady state of the output signal (within a clock phase) depends only on the value of the input signal and not on the timing.

An overview of the IKS chip system event timing model is illustrated in Figure 4-13. The box above the timing diagram shows an example of an adder module with multiple input sources. The latches are represented by filled solid rectangular boxes; different fill styles are used to denote the different control clock phases. The two block diagrams next to the box illustrate the flow of data in the two phases. The effects of the clock can be viewed as turning on or off a connecting switch above the corresponding latch as shown. Therefore, during phase one, data flow through the phase-one latch and the C1 block but the output from C1 is blocked by the open state of the second switch. During phase two, data from the C1 block flow through the phase-two latch and the C2 block, but the output from C2 is similarly blocked. Note that the adder module shows a feedback path in its block diagram, but this feedback path is nonexistent as far as each individual clock phase is concerned.

The instruction execution implements a pipeline design to achieve a higher performance. As shown in the diagram, the pipelined execution of the clock i instruction spans two clocks/four phases. In phase one of clock $i-1$, the clock i

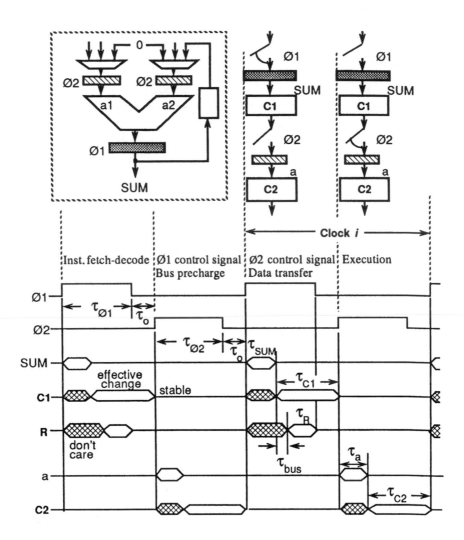

Figure 4-13. Timing model of system events with level-sensitive design.

instruction is fetched and decoded. In phase two of clock $i-1$, the bus is precharged and the control signals related to phase-one events start to propagate. During phase one of clock i, data are transferred from various sources to designations under the phase-one control signals. Meanwhile, control signals related to phase-two events start to propagate. During phase-two of clock i, the designated operations are executed. Note that the transfer of the execution results is considered part of the data transfer activities of the next instruction. More detailed timing information for the phase-one events, phase-two events, the cordic operation, and the multiplier pipelined design are specified in the following sections.

4.3.2.2.1 Timing of Phase-One Events

Phase-one system events are of two categories: data transfer and logic operations. Figure 4-14(a) shows the timing diagram of the operation of a logic circuit module C1 and the data transfer from a source S to a destination D. From the diagram, it is obvious that in order for the destination to latch the data, the phase one clock width must satisfy

$$\tau_{\varnothing 1} \geq \tau_S + \tau_{bus} + \tau_D , \qquad (4.1)$$

where τ_S, τ_{bus}, and τ_D are the delay time of the source latch, bus, and destination latch, respectively. On the other hand, the logic circuits that take effect in phase one can still experience change after the phase-one clock is

(a) Timing diagram for phase-one events.

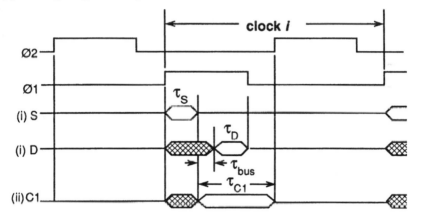

(b) Propagation of control signals (K) related to phase-one events.

    (i) Bus data transfer

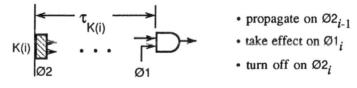

- propagate on $\varnothing 2_{i-1}$
- take effect on $\varnothing 1_i$
- turn off on $\varnothing 2_i$

    (ii) Phase-one logic circuit

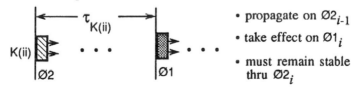

- propagate on $\varnothing 2_{i-1}$
- take effect on $\varnothing 1_i$
- must remain stable thru $\varnothing 2_i$

Figure 4-14. Timing model of phase-one events.

dropped, but they should be stable before the phase-two clock rises. Therefore,

$$\tau_{\varnothing 1} + \tau_o \geq \tau_S + \tau_{C1} , \qquad (4.2)$$

where τ_{C1} is the delay time of the logic circuit C1 and τ_o is the non-overlapped time between phase two and phase one.

The effects of this timing model on the control signals are illustrated in Figure 4-14(b). As shown in the figure, control signals that affect phase-one events must be stable before the phase-one clock rises and they may thus start to propagate during phase-two of the preceding clock. However, signals that control data transfer (K(i)) must be turned off in phase-two of the current clock as the buses start to precharge, while signals that control logic circuit functions (K(ii)) must remain stable through the entire phase-two.

4.3.2.2.2 Timing of Phase-Two Events

Phase-two system events are exclusively operations of logic circuits; the associated timing diagram is shown in Figure 4-15. In this case, the phase two clock width needs only to satisfy

$$\tau_{\varnothing 2} \geq \tau_S , \qquad (4.3)$$

where τ_S is the delay time of the operand latch. As in the phase-one case, the phase-two logic circuits delay time must satisfy

$$\tau_{\varnothing 2} + \tau_o \geq \tau_S + \tau_{C2} . \qquad (4.4)$$

In contrast to phase-one logic, phase-two logic circuits usually have multiple sources. Therefore, the signals controlling the phase-two events can be divided into two kinds: operand selection (K(iii)) and operation selection (K(iv)). The effects of the timing model on these control signals are illustrated in Figure 4-16. As shown in the figure, both of these control signals should be stable before the phase-two clock rises and they may start to propagate during phase-one of the current clock. Signals that only control operand selection need

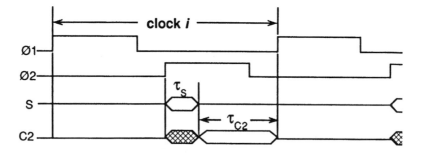

Figure 4-15. Timing diagram of phase-two events.

(iii) Operand control

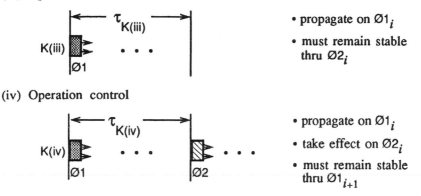

• propagate on $\varnothing1_i$

• must remain stable thru $\varnothing2_i$

(iv) Operation control

• propagate on $\varnothing1_i$

• take effect on $\varnothing2_i$

• must remain stable thru $\varnothing1_{i+1}$

Figure 4-16. Propagation of control signals related to phase-two events.

to remain stable through phase-two of the current clock, but signals that control circuit functions must remain stable through phase-one of the next clock.

4.3.2.2.3 Timing of Cordic Execution

The cordic operation can be implemented in software or hardware. With the software approach, the basic control signals needed for the iterations are stored as a subroutine in a separate location in the control store and are executed when needed. This requires hardware support of branching and returning in the instruction address control. If the subroutine is stored using a separate addressing space, then a separate counter is needed. If it is stored in the addressing space of the IKS program instructions, then not only is the addressing space available to the IKS programming decreased, but also the program counter design is more complicated. It is noted from the dataflow table that the basic control signals for establishing the operand routings are the same and the memory addressing, operand shifting, and iteration counting can all share a common counter. Hence, if the preprocessing of the operations and postprocessing of the results are separately implemented, then the hardware implementation of a cordic core just executing the iteration parts can be very simple. As a result, a hardware approach is adopted.

Figure 4-17 shows the timing diagram of the cordic execution. Signals labelled pc and cnt are the outputs of the program counter and the cordic counter, respectively. In the control word, a two-bit field called *mode* is designated for initiating the control transfer from the main control unit to the cordic control unit. A nonzero value in this field indicates a cordic operation. These two bits are fetched from the control store in phase-one and start to propagate in phase-two. Once the nonzero condition is detected by the cordic control unit, a signal (*cordic*) is raised to high to stop the program counter by blocking the incoming clocking signal. In the mean time, it enables the cordic

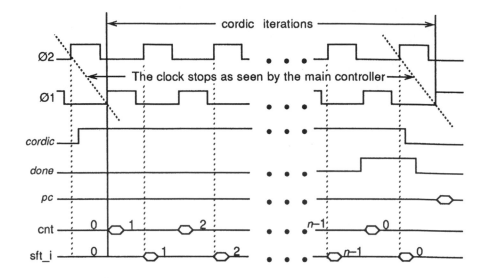

Figure 4-17. Timing diagram of the cordic execution.

counter to start counting. Latches controlled by the appropriate phase are used to latch the cordic counter output for addressing the ROM and shifting the results of the X and Y adders. When the iteration finishes, a *done* signal is raised to high, which unblocks the clocking of the program counter and resets the cordic counter to zero. If the next instruction is a normal operation, the signal *cordic* will drop and the interactions between the main control and the cordic control will return to the normal state.

4.3.2.2.4 Timing of the Multiplier

From the timing model of system events, it is clear that the clock rate, one of the major factors that affect the performance, is determined partially by the delay times of the operation modules. Using data from the LSI 1.5-μm gate array library, a 32-bit adder has a nominal delay time of 23.5 ns, while a 16-by-16 multiplier has a nominal delay time of 35 ns [LSI87b]. A 24-by-24 multiplier is not directly available from the library but can be generated from a module generator. Furthermore, the layout of the generator result can be partitioned into pipelined operations [Hin88]. Another estimate according to the 1.5-μm Cell Compiler Library from VLSI technology places the 24-by-24 multiplier delay time in the neighborhood of 80 ns [VTI88]. It is desirable to partition the multiplier logic such that the clock width of phase-two can be set to approximately the same delay time as the adders. While the physical design of the multiplier is not within the scope of this work, it is necessary to specify an achievable requirement for evaluation purposes. The multiplier timing is considered from this perspective.

Figure 4-18 illustrates the multiplier timing assuming a 2-stage pipelined design. Note that since the link from the stage-1 result to the Latch-1 is dedicated, the stage-1 delay time can be extended to the first portion of the phase-one. In most two-phase clock designs, the phase-one clock width is equal to the phase-two clock width [Sho88]. It, therefore, appears reasonable to expect that even a 80-ns multiplier can be partitioned into two stages with the clock width close to that of the adders.

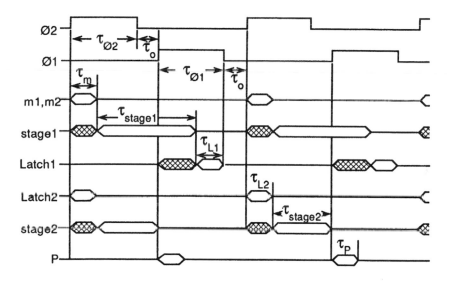

Figure 4-18. Timing diagram of the two-stage pipelined multiplier.

4.3.2.2.5 Timing Constraints

The timing constraints discussed in the previous sections are collected and classified into three categories as shown in Figure 4-19. To simplify the picture, all the latches are assumed to have equal delay time. Obviously, the clock rate is determined by the critical paths of the two phases. Guided by the specified timing constraints, building block circuits with different area-time tradeoff characteristics can be selected during the physical design phase.

4.3.2.3 The Interconnection Scheme of the IKS Chip

Major operation modules used in the IKS chip include two shifters, three adders, and one multiplier. To keep the adders and the multiplier busy all the time, resources must be provided to transfer three data items (two operands and one result) in and out of each of these operation modules in every cycle. Due to the data dependency as well as the nature of the coprocessor concept, however, such ideal parallelism does not exist. Therefore, the goal of the interconnection for the IKS chip is to seek a scheme that balances various concerns about the

1. Phase-one events related:

$$\tau_{\emptyset 1} + \tau_o \geq \tau_{latch} + \max(\tau_{C1}) \qquad (4.5)$$

$$\tau_{\emptyset 1} \geq \tau_{bus} + 2\tau_{latch} \qquad (4.6)$$

2. Phase-two events related:

$$\tau_{\emptyset 2} \geq \tau_{latch} \qquad (4.7)$$

$$\tau_{\emptyset 2} + \tau_o \geq \tau_{latch} + \max(\tau_{C2}) \qquad (4.8)$$

$$\tau_{\emptyset 2} + \tau_o \geq \tau_{MPY1} + 2\tau_{latch} - \tau_{\emptyset 1} \qquad (4.9)$$

3. Control signals related:

$$\tau_{\emptyset 2} + \tau_o \geq \max(\tau_{K(i)}, \tau_{K(ii)}) \qquad (4.10)$$

$$\tau_{\emptyset 1} + \tau_o \geq \max(\tau_{K(iii)}, \tau_{K(iv)}) \qquad (4.11)$$

$$\tau_{\emptyset 1} + \tau_o \geq \tau_{latch} + \tau_{decode} \qquad (4.12)$$

Figure 4-19. Timing constraints on the data and control signals.

utilization of the interconnection, the utilization of a dataflow pattern (and thus, an instruction), the difficulty and performance implications in circuit design and implementation, and the total computation time.

The three adders and the multiplier require a total of 12 interconnection lines for their operands and results. If all these operation modules are to have equal access to the three storage modules — the two register files and the ROM — then a bus-based interconnection scheme is the most sensible solution. With this in mind, the dataflow pattern is examined to determine the number of busses needed and the way in which these operation modules and storage modules are to be connected.

The initial dataflow table clearly shows two major distinct dataflow patterns. One is due to the cordic operation and the other is due to the computation of the type $ab + cd$. (These two patterns can be found in Figure 4-12.) During the cordic iteration steps in cycle 91, the X and Y adders' results are fed back to both adders with each having its own result and the other adder's shifted result as inputs. The Z adder result is also fed back to the Z adder input, but the Z adder needs another operand from the ROM. The multiplier is not used in the cycle. In cycles 85 and 86, the multiplier receives operands from the J and R registers, and the results are accumulated after a two-cycle delay due to the multiplier pipelining. In this case, the X and Y adders are not used. These two dataflow patterns suggest the use of local direct links for the X and Y adders, and for the multiplier and the Z adder. Furthermore, because of its frequent use and relatively low cost, the operand 0 is hardwired to relevant inputs through multiplexers. Thus, a two-bus system is sufficient to transfer the data from storage modules to the operation modules during operations involving 0's.

Because of the pipelining operation, it is desirable to have the results of the Z adder stored without using the bus so that the MAC structure can achieve its maximum throughput. Accordingly, a direct link is added between the Z adder output and the R register. On the other hand, the results from the X and Y adders can be coupled with other operations, and therefore, are simply connected to the two busses.

With this initial interconnection scheme, the dataflow table is modified to display the new dataflow pattern. Since each connection to the bus must be controlled, reducing the number of connections can reduce the demands on control resources. Accordingly, manipulations are made to reduce the need for the storage modules and operation modules to access both busses. Also, with room in the microprogram space, small increases in computation cycles are traded for other resource reductions. Specifically, to reduce the loading effects, if the number of destinations of a data transfer is more than two, then a cycle is inserted so that a certain operation is delayed. The utilization of the interconnection and the routing patterns further reveal that the Y adder output is seldom used. Consequently, the interconnection between the Y adder output and the bus is removed; and the result of the Y adder is transferred to the bus via the X adder one cycle later.

The final interconnection scheme for the IKS chip is shown in Figure 4-20. Note that there are two direct links between modules: one from the multiplier to the Z adder input and one from the Z adder output to the R register file. The local direct links between modules obviously have placement implications in physical design.

4.3.3 Design Decisions on the Control Structure and Mechanisms

The IKS execution requires less than 128 instruction cycles as found by the final IKS dataflow table. A simple on-chip control store can be implemented to accommodate the control signals with a very modest area requirement. The MACC control signals can be classified into four categories of routing control, phase-one operation, operand selection, and phase-two operation. Each category has its own timing characteristics as specified in Sections 4.3.2.2.1. and 4.3.2.2.2. The definitions of all the control signals are tabulated in Appendix C. The collection of the control signals forms the control vector of the MACC. The behavior of the MACC, represented as the MACC instruction, is the high-level interpretation of system events due to the control vector values. As defined by the timing model, each MACC instruction governs the chip's behavior for a duration of one clock cycle. The collection of all valid control vector values of the MACC forms the MACC instruction set.

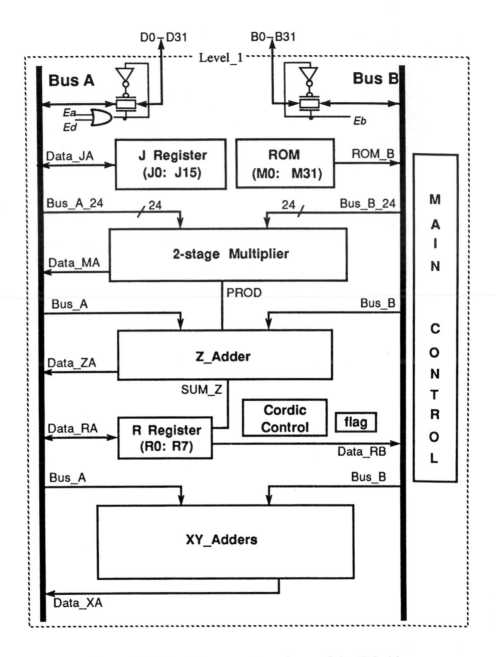

Figure 4-20. The interconnection scheme of the IKS chip.

As pointed out previously, new opportunities have been created in the design of instruction sets for algorithm-specific processors. The intended functional behavior of the MACC is basically the result of a specific sequence of instructions or control vectors. This control vector sequence creates a pattern, which may allow hardware resources for the control to be more efficiently utilized in a certain way. To explore this opportunity, the entries of the IKS dataflow table are transformed and sorted to form a symbolic control signal pattern profile. Based on the analysis of the control signal patterns, alternative encoding schemes are considered and specified. The control signals, encoded as microcodes, are then recast as the MACC instruction set. Finally, the MACC microprogram for the IKS computation is generated for the control store.

4.3.3.1 The IKS Chip Symbolic Control Signal Pattern Profile

By specifying the contents of operation modules and the operand transfer routes at each cycle, the IKS dataflow table has implicitly defined the needed control signals, which will in turn create the system events of the dataflow. These control signals, however, are obscured by other information and thus the pattern is not easy to see. To make the control signal patterns explicit, system events for each cycle specified in the IKS dataflow table are rearranged such that phase-one and phase-two events are separated. In each pattern, symbolic names are retained to represent the system events, and thus the implied control signal values, so that consistency checks can be performed easily. For example, instead of identifying the sources of a bus as 1, 2, 3, or 4, the symbolic names of X, Y, Z, P, R, J, and M are used to denote the adders and multiplier results, registers, or the ROM. Addresses of the ROM and registers are stripped off because they are treated as parameters of the instruction and are thus irrelevant. Only unique patterns are collected and grouped, and the instruction cycles where these patterns occur are indexed. This enables "lonely" patterns (patterns that are used only once or twice) to be quickly identified. With the help of the dataflow table, analysis is made to see whether lonely patterns can be eliminated or merged with other patterns without compromising performance.

The final version of the MACC symbolic control pattern profile is shown in Table 4-4. The column of phase-one patterns has 8 fields for specifying the sources and destinations of the two buses, the read/write operations of the J and R registers, the mode signal, and the operation of the Z_Angle block (see Section 5.4.3.2). When data is written to the R register the source information (bus A or the Z adder) must be specified (see Section 5.4.3.7). The column for the phase-two pattern also has eight fields: six are for specifying the operands of the three adders; one for specifying the adder operations; and one for specifying how the flag is set (see Section 5.4.3.4).

4.3.3.2 Coding Scheme Alternatives

The number of bits in the control signal pattern is 26 (not including the Register/Memory addresses). Since there are only 42 control patterns, each

Table 4-4. The MACC Symbolic Control Signal Pattern Profile.

no.	Phase_one bus A S	D	bus B S	D	Registers J	R	z a n g	z1	z2	x1	x2	y1	y2	Add (Z,X,Y)	s e t f
0															0
1	Z	z1	M	z2				A	B	X	Y	Y	X	c.c.*	0
2								0	Z	0	Y	0	X	+,+,+	2
3								0	0	0	Y	0	X	x,f,f	0
4						w-Z		P	0			Y	0	+,X,+	0
5	J	y2			r							0	A	x,x,+	1
6	J	y2			r			0	0	0	Y	0	A	x,f,f	0
7	J	z1			r		0	A	Z			Y	0	+,X,+	0
8	Z	y2						P	0			0	A	+,X,+	0
9	Z	y2								0	Y	0	A	X,+,+	1
10	Z	y2						0	0	0	Y	0	A	x,f,f	0
11							0	P	Z					+,X,X	
12	R	y2				r	0	P	Z			0	A	-,X,+	1
13	R	y2				r	0	P	Z			0	A	+,X,+	1
14	R	y2				r		0	0	0	Y	0	A	x,f,f	0
15	R	y2				r	0	P	Z			Y	A	+,X,+	
16	J	R			r	w-A		P	0	0	Y	0	X	+,f,f	0
17	X	R	R	z2		r,w-A		0	B	0	Y			-,+,X	
18	J	m1	R	m2	r	r		P	0					+,X,X	
19	J	m1	R	m2	r	r	0	0	Z					+,X,X	
20	J	m1	R	m2	r	r	0	P	Z	X	Y			+,+,X	
21	J	m1	R	m2	r	r	0	P	Z					-,X,X	
22	J	m1	R	m2	r	r,w-Z		P	0					+,X,X	
23	R	m1	R	m2		r,r	0	P	Z					-,X,X	
24	R	m1	R	m2		r,r	0	P	Z					+,X,X	
25	R	m1	R	m2		r,r		P	0	X	Y			+,+,X	
26	R	m1	R	m2		r,r	0	0	Z					+,X,X	
27							0	0	Z					-,X,X	
28							1	0	ang					-,X,X	
29	X	m1	M	m2			1	0	ang					+,X,X	0
30	Z	z1	M	z2				A	B					+,X,X	
31	Z	z1	M	z2				A	B					-,X,X	
32	Z	z1	M	z2		w-Z		A	B					-,X,X	
33	R	z1	M	z2		r		A	B					+,X,X	
34	J	z1	M	z2	r	w-Z		A	B					-,X,X	
35	J	z1	M	z2	r	w-Z		A	B					+,X,X	
36	R	z1	R	z2		r,r	0	A	B					-,X,X	0
37	R	y2	M	x2		r	0	0	Z	0	B	0	A	+,+,+	0
38	R	y2	M	x2		r	0	0	Z	0	B	0	A	+,+,+	2
39	P	y2	M	x2						0	B	0	A	X,+,+	1
40	X	R	M	x2		w-A				0	B	Y	0	X,+,+	
41	Z	J	M	x2	w		3	0	ang	0	B	0	0	+,+,+	3
42	Z	J	M	x2	w		1	0	ang	0	B	0	0	+,+,+	0

* The adder control signals are taken from the cordic control unit.

pattern can be represented uniquely by a 6-bit opcode. It is desirable to thus store only the opcode rather than the entire pattern. But the timing of the instruction fetch (including a counter delay and memory access delay) and the opcode decoding must satisfy the timing constraints of eq. (4.12). Two alternatives that do not require changing the timing model of instruction execution are considered here.

A brute-force encoding scheme is to use a 6-bit number to encode the signal patterns. This scheme is simple and the storage area required for the opcode is minimal. Also, since a 6-bit number can encode up to 64 patterns, the unused coding space allows the addition of new patterns to either speedup the computation or facilitate testing. The disadvantage of this approach is that the decoding time may be too long in the overall time budget of the instruction fetch/decode phase.

The second scheme decomposes the signal pattern into smaller subpatterns and encodes each subpattern separately. Because of the smaller size, each subpattern can be decoded faster. This scheme, however, may require a larger storage area.

Essentially, the two schemes represent a tradeoff between area and time. To determine which scheme is adopted for the IKS chip instruction decoding, rough delay times for various phase-one event cases are estimated based on the gate delays of major components. Comparison of the delay time estimates indicates that the phase-one clock width may well depend on the entire instruction fetch/decode process. In order to achieve a faster clock rate, the second scheme is chosen.

Observation of the control signal pattern profile reveals that a number of patterns differ only in adder functions. Separating the adder function encoding results in two opcode fields. The number of patterns for the larger field is reduced to 29, which can be encoded in a 5-bit number. The adder operation field can be encoded with a 3-bit opcode. This encoding scheme, together with the codemap specifications is given in Appendix D.

4.3.3.3 The MACC Microcode and Instruction Set

With the encoding scheme determined, the two opcode fields are combined with the address fields to form the MACC microcode. Figure 4-21 shows the format of the MACC microcode and the entire control vector format after decoding. Because ROM addressing and the R register read (via Port B) never occur concurrently, they can share the same address field. The two opcode fields are decoded into two vectors of 20-bit and 6-bit wide. The first 11 bits for the 20-bit vector are decoded from opcode 1. These signals control phase-one events (*ctrl_v1*). The remaining bits are concatenated with the 6-bit vector decoded from opcode 2 to form the phase-two events control vector (*ctrl_v2*).

To facilitate documentation and testing, the control signal patterns are recast as the MACC instruction set by specifying their microcode values and the corresponding behavioral interpretation of the resulting system events. The

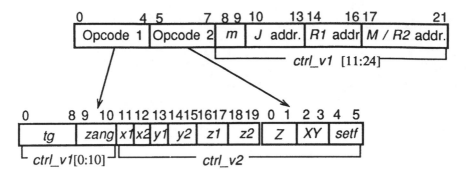

Figure 4-21. The MACC microcode format and the control signals.

MACC instruction set is shown in Table 4-5. Each table entry specifies an MACC instruction number, the microcode, and the operations. The microcode consists of two opcode numbers and four parameters — one for the mode control, and three for addresses. Only address fields required by the instruction execution are indicated in the address parameter column. RA and RB indicate Port A and Port B addresses of the R register, respectively. A square bracket denotes that the address items inside will be used in actual instruction execution but they have no effect on the functional results. In the operation field, X, Y, and Z denote the three adder results, and P denotes the multiplier result. A prime post-superscript denotes the result of the last cycle, and a double prime pre-superscript in P denotes results delayed by two cycles due to the multiplier pipelining. J and M stand for the contents of J register and ROM specified by their addresses, respectively. The function of $ang(Z')$ and $ang(Z',f)$ are operations of the Z_Angle module (see Section 5.4.3.2).

4.3.3.4 The MACC Microprogram for the IKS Computation

In the MACC microcode and instruction set table, the index field contains instruction cycle information for the IKS execution. The instruction set table is first expanded such that each entry indexes to exactly one instruction cycle. The entire table is then sorted using the index as the key. Values of the address fields are specified by tracing back to the corresponding cycle in the IKS dataflow table via the index. The entire MACC microprogram for IKS execution is listed in Appendix E.

4.3.4 Evaluation

The effectiveness of the IKS chip architecture is evaluated in terms of area, speed, and resource utilization. Since the actual performance depends on the actual physical design, the goal of this evaluation is to provide bounds on these categories. In addition, testability considerations are addressed qualitatively.

4.3.4.1 Area

The IKS chip design follows the top-down hierarchical approach. With this approach, the architecture is (structurally) decomposed until its building block

Table 4-5. The MACC Instruction Set.

Inst. no.	Opcode path	func	Parameters m	Address	Operation
0	0	0	0		Z=0. X=Y', Y=X'
1	1	1	>0		Z=Z'(Az)M, X=X'(Ax)sft[Y'], Y=Y'(Ay)sft[X']
2	29	6	0		Z=Z', X=Y', Y=X', f=sign[Z']
3	0	3	0		Z=0, X=(f)Y', Y=(f)X'
4	2	0	0	RA	Z=P', X=Y', Y=Y', RA←Z'
5	3	4	0	J	Z=0, X=Y', Y=J, f=sign[y2]
6	3	3	0	J	Z=0, X=(f)Y', Y=(f)J
7	4	0	0	J	Z=J+Z', X=Y', Y=Y'+J
8	5	0	0		Z=P', X=Y', Y=Z'
9	6	4	0		Z=0, X=Y', Y=Z', f=sign[y2]
10	6	3	0		Z=0, X=(f)Y', Y=(f)Z'
11	7	0	0	[RA]	Z=P'+Z', X=Y', Y=RA
12	7	5	0	RA	Z=P'−Z', X=Y', Y=RA, f=sign[y2]
13	7	4	0	RA	Z=P'+Z', X=Y', Y=RA, f=sign[y2]
14	8	3	0	RA	Z=0, X=(f)Y', Y=(f)RA
15	9	0	0	RA	Z=P'+Z', X=Y', Y=Y'+RA
16	10	3	0	J,RA	Z=P', X=(f)Y', Y=(f)X', RA←J
17	11	0	0	RA,RB	Z=−RB, X=Y', Y=0, RA←X'
18	12	0	0	J,RB	"P=J*RB, Z=P', X=Y', Y=0
19	13	0	0	J,RB	"P=J*RB, Z=Z', X=Y', Y=0
20	14	0	0	J,RB	"P=J*RB, Z=P'+Z', X=X'+Y', Y=0
21	14	2	0	J,RB	"P=J*RB, Z=P'−Z', X=X'+Y', Y=0
22	15	0	0	J,RA,RB	"P=J*RB, Z=P', X=Y', Y=0, RA←Z'
23	16	2	0	RA,RB	"P=RA*RB, Z=P'−Z', X=Y', Y=0
24	16	0	0	RA,RB	"P=RA*RB, Z=P'+Z', X=Y', Y=0
25	17	0	0	RA,RB	"P=RA*RB, Z=P', X=X'+Y', Y=0
26	18	0	0	RA,RB	"P=RA*RB, Z=Z', X=Y', Y=0
27	18	2	0	[RA,RB]	Z=−Z', X=Y', Y=0
28	19	2	0	[M]	Z=−ang[Z',f], X=Y', Y=0
29	19	0	0	M	"P=X'*M, Z=ang[Z',f], X=Y', Y=0
30	1	0	0	M	Z=Z'+M, X=X'+Y', Y=Y'+X'
31	1	2	0	M	Z=Z'−M, X=X'+Y', Y=Y'+X'
32	20	2	0	RA,M	Z=Z'-M, X=Y', Y=0, RA←Z'
33	21	0	0	RA,M	Z=RA+M, X=Y', Y=0
34	22	2	0	J,RA,M	Z=J−M, X=Y', Y=0, RA←Z'
35	22	0	0	J,RA,M	Z=J+M, X=Y', Y=0, RA←Z'
36	23	2	0	RA,RB	Z=RA−RB, X=Y', Y=0
37	24	0	0	RA,M	Z=Z', X=M, Y=RA
38	24	6	0	RA,M	Z=Z', X=M, Y=RA, f=sign[Z']
39	25	4	0	M	Z=0, X=M, Y=P' if f=0; Y=0 if f=1
40	26	0	0	RA,M	Z=0, X=M, Y=Y', RA←X'
41	27	7	0	J,M	Z=ang[Z'], X=M, Y=0, J←Z'
42	28	0	0	J,M	Z=ang[Z',f], X=M, Y=0, J←Z'

circuits can be defined in terms of some commonly accepted circuit primitives such as logic gates, multiplexers, and latches. Area requirements of these hardware primitives are, in most cases, estimated based on the effective gate counts according to LSI Logic's 1.5-µm gate array library [LSI87a]. The area of the 24-by-24-bit multiplier is extrapolated from the data of LSI Logic's multiplier macrofunctions [LSI87b]. For the areas of PLAs (Programmable Logic Arrays) within the main control module, the estimates are based on information from the databook of VLSI Technology's Macrocell Compiler [VTI88]. The areas of the higher level modules are obtained by adding up the effective gate counts of the component modules and the sums are divided by a utilization factor. When a design is decomposed hierarchically, design regularity tends to increase, and as a result, the utilization factor tends to be higher [LSI87a]. Accordingly, a utilization factor of 0.8 is assumed for the functional modules to provide the estimate of practical gate count. Detailed gate/area estimates of each module can be found in [Leu89a].

Table 4-6 shows the effective and practical gate counts and the area percentage of each module. Note that if the area taken by the storage modules and the control units are roughly invariant in all architectures implementing the IKS algorithm, then the addition of a multiplier to a cordic core only increases the total chip area by 23.3 percent. Looking at it another way, the addition of two adders (the XY_Adders module) to a MAC only results in an increase of 30.5 percent in total chip area. So the merger of the MAC and cordic in this design appears economical in area when compared with a SIC coprocessor

Table 4-6. Estimates of the IKS Chip's Effective and Practical Gate Counts.

Module	Gate (eff.)	Utilization	Gate (prac.)	Percentage
Multiplier	5288	0.8	6610	18.88%
Z_Adder	2809	0.8	3511	10.03%
XY_Adders	6548	0.8	8185	23.37%
Set_Flag	17	0.8	21	0.06%
ROM	427	0.8	534	1.52%
J Register	5857	0.8	7321	20.91%
R Register	3936	0.8	4920	14.05%
Main Control	2999	0.8	3749	10.71%
Cordic Control	131	0.8	164	0.47%
Total	28013	0.8	35015	100.0%
		0.7	30017	
		0.6	46687	

approach. Also shown in the table are the total gate count and area of the IKS chip with different utilization estimates. Current commercial gate array technology with a feature size of 1.5 μm can provide up to 50k usable gates in a single chip [LSI87a]. From these figures, it can be seen that the entire IKS chip can be fabricated on one chip even for a gate utilization as low as 60 percent.

4.3.4.2 Performance

The IKS execution by the MACC (not including I/O) takes 445 clock cycles. According to the system timing model, the width of the phase-two clock is about the duration of an adder delay. As the delay time of the 32-bit adder from LSI Logic is less than 25 ns, a 100-ns clock cycle appears more than achievable. This leads to a lower bound estimate of 45-μs computation time. With a more aggressive clocking scheme, it may be possible to reduce the clock cycle by 30 percent. This clock rate gives an estimate of a computation time of about 30 μs.

These two estimates compare very favorably with previous designs such as Lee and Chang's maximum pipelined design, which achieves a throughput rate of 40 μs using 25 cordic processors. The 40-μs figure in their design, however, is estimated based on earlier technology. It actually represents one cordic cycle. To make the comparison more meaningful, the speedup of both designs are compared to the execution cycles on a single cordic processor. It has been shown that the single cordic processor execution of the IKS algorithm takes 25 cordic cycles [LeCh87]. The maximum pipelined design has a latency of 18 cordic cycles, i.e., a speedup factor of 1.39. In the MACC design, one cordic cycle takes an average of 27 cycles, (1 for preprocessing, 24 for iterations, and 2 for postprocessing,) and thus the total computation time of 445 cycles is equivalent to 16.48 cordic cycles, i.e., a speedup of 1.517. In the last section, the area of the MACC is estimated to be 1.233 times that of a single cordic processor. Using AT^2 as the figure of merit for estimating the area-time efficiency of an architecture/algorithm design, the MACC is 1.866 times more efficient than a single cordic processor in computing the IKS.

4.3.4.3 Resource Utilization

The utilization of a hardware resource is defined as the ratio between the number of cycles that the resource is actually used and the total number of cycles that the IKS computation takes. Since cycles can refer to instruction cycles or clock cycles, the utilization can be viewed similarly. The main difference between these two categories is that a cordic operation is counted as one instruction cycle but requires 24 clock cycles. Table 4-7 shows the utilization of operation modules and busses in the two categories.

Viewed at the higher level of instruction cycle, the utilization of these resources are unimpressive. But when viewed at the level of clock cycle, the

Table 4-7. Hardware Resource Utilization of the IKS Chip.

Resource	I-cycles (%)	Clock (%)
Bus A	94 (75.2)	370 (83.15)
Bus B	68 (54.4)	344 (77.3)
Multiplier	35 (28.0)	35 (7.86)
Z adder	82 (65.6)	358 (80.4)
XY adders	52 (41.6)	374 (84.0)

level at which the computation is carried out, the cordic core (the three adders and the two busses) achieves an average utilization of 80%. It is interesting to note that the multiplier is used only 7.8% of the time, but it reduces the computation time by one third.

Another utilization evaluation involves the control resources. The 7-bit program counter provides a program space of 128 instruction addresses. The IKS computation takes up 125 instruction cycles indicating a 97.6% utilization. The 5- and 3-bit decoding spaces of the two opcode fields achieve 90.6% and 100% utilization, respectively. The instruction set is 100% useful as expected since it is derived from the algorithm. The utilization profile of the instruction set is shown in Table 4-8. Here it is seen that four instructions, or less than 10% of the instruction set, account for 32.8% of all instructions used in the IKS microprogram. On the other hand, as many as 20 instructions, or 46.5% of the instruction set, are used only once in the microprogram. This highly customized feature of the instruction set, as a result of engineering various hardware resources to facilitate the dataflow of the algorithm, together with other

Table 4-8. Utilization of the MACC Instruction Set
in the IKS Calculation.

Frequency	Number of instructions	Total usage (%)
1	20	20 (16.0)
2	5	10 (8.0)
3	7	21 (16.8)
4	4	16 (12.8)
5	1	5 (4.0)
6	2	12 (9.6)
8	1	8 (6.4)
9	1	9 (7.2)
10	1	10 (8.0)
14	1	14 (11.2)

decisions on resource allocation, contributes to the MACC's excellent effectiveness in computing the IKS.

4.3.4.4 Testability Considerations

At the beginning of this design task, it was not completely clear what kind of functional units should be used, let alone how they would be tested. Therefore, the initial stages of the IKS chip design could only follow some general guidelines on design for testability. These guidelines include [Hna87]:

- Design initializable circuits;
- Avoid redundancy;
- Avoid race conditions;
- Avoid asynchronous timing;
- Provide access to device clocks;
- Allow feedback to be open/closed;
- Provide access to major busses;
- Use wired logic cautiously;
- Use control and test points;
- Use circuit partitioning and selective control.

Once the functional units are determined and the interconnection scheme is developed, the conditions for fine-tuning the approach of design for testability become ripe. Because of the bus-oriented design and the hierarchical decomposition along the functional line, each functional module in the MACC datapath is self-contained and has clear functional definitions. Since these modules are built on vendor-supplied library modules, well proven test data usually has been developed and can be used. It is therefore determined that the goal of testability for the IKS chip design is to enable the individual testing of these functional circuit modules.

One overriding consideration in adopting a particular testing strategy is the chip's I/O capability. The gate arrays that can provide the required gate counts for the IKS chip typically have more than 100 I/O pins. But because of the simplicity of the I/O in the IKS, many pins will be left unused. Since the functional modules of the MACC can be accessed via the two busses, unused I/O pins can be used to directly assess the busses. This will speedup the data transfer during testing.

Since the input operands and output results of the functional modules can be observed directly via the two busses, scan capability is not necessary for the internal latches of the modules but level-sensitive design is still required. The only latch that must have scan in/out capability is the control signal latch from the control store. To speedup the testing of the register files, latches that have reset capabilities are selected as the basic cells to form the registers. The detailed designs of the module circuits are then reexamined for data transfer conditions during testing. The only modification needed, as the result of the reexamination, is the addition of a few multiplexers in the XY_Adders module. The details are described in Section 5.4.3.3.

4.4 Summary

An architecture for computing the closed-form solution of the inverse kinematics of a robot manipulator has been described. The architecture, called MACC, is based on the concept of embedding a cordic core in an MAC structure, or looking it another way, incorporating a multiplier coprocessor into a cordic core. The entire IKS computation requires 125 instructions, or 445 clock cycles. The effective gate count of the chip is about 28k, and the practical gate count is estimated to be less than 50k even for a poor gate utilization of 60%. Based on current gate array technology, the architecture can thus be implemented in a single chip. If a clock rate of 10 MHz is achieved, the latency of the IKS computation is less than 45 µs. Compared with a single cordic processor implementation, the area of the MACC is estimated to be about 23% larger, but the computation time is reduced by more that one third, thus giving a overall area-time efficiency of 1.866 over the single cordic processor approach.

Major design decisions regarding the IKS chip design have been documented. Furthermore, the process of making these design decisions has been generalized into an ASIC architecture design methodology. Based on the conceptual framework for ASIC design, the methodology partitions the architecture design process into three phases of functional unit configuration, communication configuration, and the control configuration. For each design phase, the methodology provides a focus for recognizing the interactions between the algorithm characteristics, architecture styles, and the implementing technology. Insights on these interactions are generalized into decision guidelines. As an investigating vehicle, the design of the IKS chip has led to the development of a number of representations that make essential aspects or features relevant to a particular design phase explicit and conducive to manipulation.

The underlying principle of the ASIC architecture design methodology is that hardware resource allocation decisions must be based on the tradeoff analysis of the applications needs and the potential benefits. In the algorithm-specific processor design environment, this design principle is manifested as a design philosophy characterized by deriving the processor's instruction set from the algorithm to be implemented. In this regard, the ASIC architecture design methodology provides a systematic approach to this end as demonstrated by the derivation of the MACC instruction set. By illustrating what design decisions are involved and how they are made in this process, the IKS chip design presents a paradigm for the class of algorithm-specific processor designs.

Chapter 5

VHDL Simulation of the IKS Chip

A good simulation, be it a religious myth or scientific theory, gives us a sense of mastery over our experience.

Heinz Pagels
THE DREAMS OF REASON (1988)

5.1 Introduction

The fundamentals of VHDL (VHSIC (Very High-Speed Integrated Circuits) Hardware Description Language) are presented in this chapter. The emphasis is placed on the concepts of signal and design entity. The presentation is organized from a user's perspective viewing VHDL as a programming language, as a design tool, and as a design environment. Further detailed information on VHDL can be found in [LRM88, URG85, D&T86, Tut88, UsCL89, Arm89, Coe89].

The simulation objective of this work is then examined and the modeling approach is subsequently determined. The discussion of the modeling is focused on two crucial issues: data typing as an abstraction mechanism, and how to *manage* the circuit delay time information.

The circuit hierarchy of the IKS chip has five levels. While the design follows the top-down approach, the simulation follows a more natural bottom-up approach. If the presentation of the VHDL description, however, strictly follows either a top-down or bottom-up approach, some important aspects of either the design or the simulation may become obscured. Therefore, we opt for a compromise, which is facilitated by the language itself. Specifically, all circuit modules are organized into (VHDL) libraries according to their respective hierarchical levels. The presentation then flows according to the logical structure of the libraries.

After the functional specifications of each module are verified by the module's respective simulated behavior, the functional and storage modules are assembled to form the datapath and the remaining modules are connected to form the control section. The datapath, together with the cordic control unit, is tested for the cordic operations. The datapath is then combined with the control section to form the MACC machine. The control section is initially configured with a control store containing the MACC instruction set. With the operations of the instruction set verified, the control section is then reconfigured to hold the IKS microprogram and the functionality of the IKS chip is simulated with realistic input data. Simulation experiments on the cordic operation, the MACC instruction set, and the IKS computation are described in Section 5.5.

5.2 VHDL Fundamentals

The VHDL project was initiated by the US Department of Defense in the summer of 1981. The language development was contracted to the team of Intermetrics, IBM, and Texas Instruments. In August 1985, Intermetrics released the VHDL 7.2 Language Reference Manual, which has become the baseline language of the IEEE standard. In September 1987, after a two-year review process, the IEEE standardization committee approved the final draft of the language, which is based on a revised version of VHDL 7.2, as the IEEE Standard 1076-87.

VHDL is intended for describing hardware from the system level down to the gate level. All phases of design — hierarchical decomposition, verification, synthesis, testing, communication of data, maintenance, modification, and procurement — are supported. The language design guideline mandates that any real hardware be expressible with the language and that features of the language be hardware realizable. The implementation of the language results in a minimum environment comprised of the analyzer, the design library (managed by the design library manager), and the simulator. The environment is expected to provide a platform for integrating existing and new CAD tools for behavior-oriented designs.

As a programming language, VHDL inherited many of the latest research results in the area of imperative languages, especially through Ada. While many of its basic programming constructs and features are similar to modern imperative languages, VHDL has some language concepts that are slanted for modeling hardware structures and behavior. Among them, the most important ones are the concepts of signal and design entity.

5.2.1 Signal and Associated Concepts for Modeling Behavior

A signal is a symbolic object corresponding to a physical wire, or a group of wires. The value of a signal is an interpretation of the wire's electrical state. Since the function of a circuit, from simple to complex, is characterized by its I/O, as embodied in wires of certain kinds (a strip of metal, silicide, polysilicon, etc.), the collection of signals corresponding to these I/O wires thus constitutes a functional equivalence of the circuit in the symbolic domain. Manipulations of these symbolic objects result in the simulated behavior of the circuit. Concepts associated with signals in VHDL by and large regulate how the physical realities can be specified by the corresponding symbolic objects' manipulations that are meaningful to the simulator.

Signals resemble variables in conventional programs; they must be declared to be certain types and their values can be changed through signal assignment statements. Signals, however, are created for modeling wires exclusively and the simulation mechanism is tied to the state of the signals, but not to any variable. Associated with each signal, there is a built-in time-dimension as shown in Figure 5-1. This time dimension is defined by the language and the associated information is maintained by the simulator through drivers on an absolute scale (with the finest resolution in fs — 10^{-15} second). The task of describing the behavior of the modeled hardware thus becomes a task of manipulating this time-dimension information.

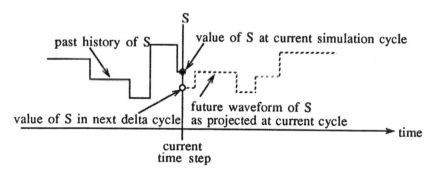

Figure 5-1. Signal S and its built-in time-dimension.

To facilitate the modeling of abstract circuit behavior, VHDL provides two delay time description primitives: **transport** delay and **inertia** delay. The target signal under the transport delay model will respond faithfully to the input signal no matter how brief the stimulus is. Under the inertia delay model, the target signal will respond to the input stimulus only if the driving event lasts for a period no less than the specified time. This timing model is useful in characterizing some circuit properties such as setup time or hold time. In addition, the language also provides predefined attributes associated with each signal. They can be used to model a wide range of high-level circuit characterizations such as hold time, setup time, transient response, edge sensitivity, etc.

5.2.2 Design Entity and Associated Concepts for Describing Structures

The design entity is regarded as the principal hardware abstraction mechanism in VHDL [LRM88]. A design entity corresponds to a hardware device of arbitrary complexity. Language concepts associated with the design entity are primarily designed for support of decomposing circuit hierarchies. They also facilitate alternative implementations, hiding of proprietary design information, and reuse of previous designs. Table 5-1 compares the two concepts of signal and design entity. Note that no matter how many levels are nested within a design entity, the behavior is ultimately realized by actions from signal assignment statements.

Table 5-1. A Comparison of the Concepts of Signal and Design Entity.

Category	Signal	Design Entity
modeled objects	wires	devices
domain of interests	behavioral: I/O relationship	structural: decomposition/connectivity
abstraction dimension	data typing, concurrency	data encapsulation
manipulation means	assignment statement	component declaration, instantiation, configuration

A design entity consists of an interface (called entity declaration) and one or more alternative bodies (called architecture). The interface establishes communication channels between the design entity and the outside world through generics and ports. The generics allow the parameterization of anything within the entity. Inside the architecture body, the values passed from the generic interface list are considered as constants. A port is the intersection point of a signal and the design entity's boundary. Thus, ports are of the same object class as signals. Certain restrictions are placed on the drivers and readers of ports to ensure the consistency of a design.

The body of an entity (via the architecture body declaration) specifies what the function of the design entity is. The architecture body descriptions can be classified into three styles: structural, behavioral, and dataflow. The structural description focuses on the connectivity between the components, and is usually the preferred style for describing the overall organization of a complex design. The behavior of the system in this style is embodied in its components. The behavioral description allows algorithmically defined input-output transformations and is usually applied to the leaf-nodes of the circuit. Behavioral descriptions can be used to specify the function of a circuit without concern as to how it is implemented. The dataflow description denotes a style between pure structural and pure behavioral. The division of these styles is artificial; the choice of a particular style for a particular design is completely under the designer's discretion.

Support for the decomposition of a circuit hierarchy, the top-down or bottom-up design approach, the reuse of previous designs, and the selection of different implementation alternatives, are among the most desirable features of VHDL. All these features depend on a language mechanism called the instantiation process, a mechanism for a design entity to use another design entity of a lower level. The instantiation process involves three steps: creating a template through *component declarations*; duplicating objects from the template through *instantiation statements*; and binding the objects to the existing (design) entities through *configuration statements*. The bindings select design units from the library for the instantiated components as well as establish the formal ports of the template and the formal ports of the design units. Individual and default bindings are available. The instantiation process forces the design entity being instantiated to be one level lower than the current level. A hierarchy of many levels can be created by extending this process from either this level up or the next level down, or both.

5.2.3 The VHDL Environment

The design environment inherently depends on implementation and is not completely defined by the language. The following description of the VHDL environment is based on the Intermetrics implementation [UMSE88]. In this implementation, the VHDL support environment consists of six components: the analyzer, the reverse analyzer, the design library, the design library manager, the simplifier, and the simulator. Figure 5-2 shows the functional

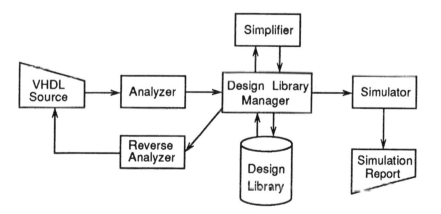

Figure 5-2. The VHDL supporting environment [Tut87].

relationship among these components. The flow of VHDL simulation is illustrated in Figure 5-3. The process of translating a VHDL hardware description into the Intermediate VHDL Attributed Notation (IVAN) format is called **analyze**. The resultant IVAN module, called a design unit, is stored in the VHDL design library and is classified into one of the five categories: entity, architecture, package declaration, package body, or configuration. The access to these design units must be through the library manager (an interactive program

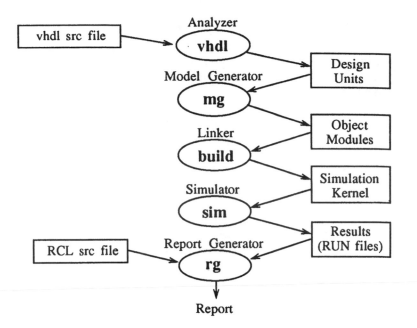

Figure 5-3. The flow of VHDL simulations.

called VLS (VHDL Library System)). The simulator consists of four programs. The model generator (**mg**) translates the IVAN modules into object code modules and places them in the simulation library. The **build** program links various object code modules corresponding to a particular design with the run-time simulation core to form the simulation kernel (a run-time image). The **sim** program executes the kernel to simulate a hardware experiment. The result of the simulation is a database file (a RUN node in the design library), which records all the transactions of the signals. The simulation results can be retrieved from the RUN node by the report generator (**rg**) program, which essentially formats the output result under the instructions contained in a Report Command Language file (denoted by a file whose filename has the extension of .RCL). All manipulations on design units/files in both the VHDL design library and the simulation library are handled through the VLS.

The control of the simulator dynamics is presented as a set of virtual test equipment called the test bench. All simulation cases must be constructed as a unit under test (UUT) submitted to the test bench for analysis. This concept is illustrated in Figure 5-4. The entity declaration of the UUT must not have signal ports, and the generics, if present, must be integer types. The UUT must be model generated with the –top option. The minimum configuration for the test bench includes the Test_Bench_Control, the Signal_Trace_Data_Recorder, and the Bed_of_Nails. The Test_Bench_Control allows the user to set the

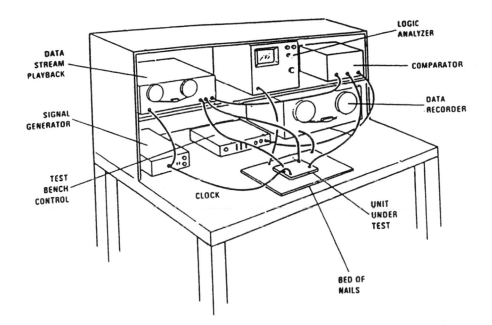

Figure 5-4. Virtual test equipment and the test bench [Tut87].

limits of the maximum simulation time, maximum delta simulation cycle, and the maximum number of transactions for each cycle. The Signal_Trace_ Data_Recorder controls the recording of signal transactions to the signal trace file. The Bed_of_Nails provides simulation information and statistics to the user. This information, available as signals, can be used to dynamically control the simulation.

5.3 Simulation Objective and Modeling Approach

Circuit simulations, in general, serve two objectives. The first is to verify that the circuit as represented by the model is indeed what is intended. The second is to investigate alternatives that may enhance, ideally in a global way, the circuit performance. Different simulation objectives require different modeling approaches, as a particular modeling approach is a compromise between the modeling costs (both human and computational) and the degree of accuracy by which a model's simulated results mimic reality. With the focus on architecture design, the simulation objectives of this work are primarily to verify the correctness of the chip's logic design, and secondarily to obtain a worst case performance estimate. The modeling approach to be discussed in this section reflects these objectives.

The validity of verification through simulation depends on two conditions: the correctness of the model in representing the relevant aspects of the circuit and how objectively the simulation process recreates reality. Confidence in using simulation as a verification means will certainly increase if efforts for modeling and for simulation are independent of each other. In this regard, simulating the IKS chip's behavior in VHDL is obviously desirable, particularly in view of the recent acceptance of VHDL by the IEEE as the standard hardware description language.

The validity of verifying the IKS chip design through VHDL simulation then depends on how the relevant aspects of the circuits are represented in the models. Two issues are involved here. They are the model semantics and the model's parameter values. The encoding of circuit behaviors in the programming semantics has four facets: *data structure, description style, data typing,* and *delay time.* Among these four facets, the data structure is the most idiosyncratic in that the use of a particular structure largely depends on the designer's perception of the circuit's characteristics.

As to the second facet, VHDL supports three architecture description styles: *structural, behavioral,* and *dataflow.* The application of these different styles is strongly related to the hierarchy level of the target circuit module. This is because in contrast to the top-down direction of circuit decomposition in the detail design, the building of models for simulation is generally a bottom-up process. Therefore, a behavioral style is applied to the bottom-level or leaf-nodes of a circuit hierarchy since the behaviors of higher level circuits are constructed from these building block circuits. The further up the hierarchy, the more convenient it becomes to describe a circuit in a more structural style. The term dataflow here refers to a description style somewhere between pure behavioral and pure structural and is usually applied to circuits at intermediate levels.

The remaining two facets of data types and delay time characterize the modeling approach of this work and are discussed in detail in the following subsections. The method of deriving parameter values is explained in the discussion of modeling the delay times.

5.3.1 Data Types

Data typing as an abstraction mechanism defines the interpretation of symbolic objects. In VHDL, all physical realities with dynamic consequences must be represented by signals in the symbolic domain [LRM88]. By defining signals of certain types, the designer extracts the relevant aspects from the real circuit for manipulation.

Aspects of the circuit can be represented at different abstract levels. Consider a two-bit signal interpreted at a more abstract level as an unsigned number. The event that the signal's value changes from 1 ("01") to 2 ("10") may actually occur as two events in sequence: from 1 to 0 ("01" to "00"), and then from 0 to 2 ("00" to "10"), as the discharge time may be shorter than the charge time. It is not difficult to see that the complexity for modeling words of more bits will increase rapidly, especially if one takes into account other factors such as different loading conditions or the effects of simultaneous switching. A more accurate model may require simulation of an object's behavior at the lower bit-level and give a word-level interpretation every time any single bit has changed its value.

Such a modeling approach requires considerably more computational resources but may not be necessary since what is relevant also depends on how the modeled object is used. Signals in the IKS chip can be classified according to usage into three categories of control, data, and address. Any change is significant for signals performing control functions as the transition may last long enough to set or reset some circuits. In addition, such signals are likely to be manipulated through Boolean functions. Therefore, control signals are essentially bit-types, Their use as BIT_VECTORs is only for convenience in transferring them between modules.

The data category includes all input operands and output results of the multiplier, adders, shifters, and memory. The signals of this category are relevant only at the word-level. The effects of transitions in these signals are generally confined to their respective functional modules and will not affect other signals, partly due to the two-phase non-overlapped clocking scheme. Furthermore, because of the level-sensitive design discipline, the value (i.e., the interpretation) of a data signal takes effect only when the signal is stable. Therefore, it is sufficient for these signals to use two's complement integers as the data types so long as: 1) the propagation delay of the data signal passing through a circuit module is no less than the worst-case delay of its component bits, and 2) the width of the clock phase is long enough such that the setup times for all the d-latches controlled by that phase are satisfied. The use of integers instead of single bits for data signals not only saves computation effort, but it also provides a vital link between the task algorithm and the architecture.

Signals that are used for memory addressing have some peculiar properties. On the one hand, they are more conveniently interpreted as integers similar to the data signals. On the other hand, unlike the data signals, transition scenarios such as that of the two-bit signal example may have harmful effects on addresses. For example, when a write operation is enabled, a fluctuation in address signals may cause the data to be written into unintended locations. To prevent the inadvertent writing of memory, the system event timing model stipulates that the read/write signals take effect only in the data transfer phase during which the address signals are guaranteed stable. In addition, inside the VHDL description of storage modules, assertion statements are incorporated to check for conditions of address changes while writing is in effect. With this protection, the address signals can be modeled directly as unsigned integers.

In addition, a special data type of *virtual_signal* is created for representing the connection types in modeling the transmission gates. The use of this data type is explained in Section 5.4.2.7. The basic data types used in the IKS chip

```
package Custom_Types is    -- Reside in Library IKS_Chip
   -- DATA
subtype BIT_32 is INTEGER;
subtype BIT_24 is INTEGER range -8388607 to 8388607;
type REG_FILE is array (INTEGER range <>) of BIT_32;
   -- CONTROL
subtype VECT_2 is BIT_VECTOR (0 to 1);
subtype PLA32_WORD is BIT_VECTOR (0 to 19);
subtype PLA8_WORD is BIT_VECTOR (0 to 5);
subtype C_VECTOR_1 is BIT_VECTOR (0 to 10);
subtype C_VECTOR_2 is BIT_VECTOR (0 to 14);
type SOURCE_2 is (src0, src1);
type SOURCE_4 is (src0, src1, src2, src3);
   -- ADDRESS
subtype BIT_7 is INTEGER range 0 to 127;
subtype BIT_5 is INTEGER range 0 to 31;
subtype BIT_4 is INTEGER range 0 to 15;
subtype BIT_3 is INTEGER range 0 to 7;
subtype BIT_2 is INTEGER range 0 to 3;
   -- VIRTUAL_SIGNAL
subtype VIRTUAL_SIGNAL is INTEGER range -1 to 2;
end Custom_Types;
```

Code 5-1. The VHDL Package Custom_Types.

description are put into the package Custom_Types under the library IKS_Chip. The declaration of package Custom_Types is listed in Code 5-1.

5.3.2 Delay Time

Major factors that affect the delay time of a circuit include capacitive loading, signal value and strength, slew rate, temperature, and coupling between signals. Among these factors, the capacitive loading is generally accepted as the primary factor. Since all these factors depend, in varying degrees, on the physical design and because the main focus of the modeling effort at this stage is to verify the correct functionality of the entire chip, only the capacitive loading factor is considered in this work. Other secondary factors can be taken into account by giving a wide enough margin to the results. (Actually, if the review of an initial design is favorable for physical implementation, architecture bodies that incorporate more accurate delay time models can be developed to replace the more global ones. This progressive simulation strategy is supported by VHDL.)

In the remainder of this section, a general delay time model using the timing description primitives provided by VHDL is described. This is followed by an explanation of how the timing parameter values are derived. Determination of certain factors, such as the fanout or interconnect length, that would affect the timing parameter values may not be possible until the overall design takes shape. A strategy based on the VHDL construct of *configuration* is developed to postpone the specification of these parameter values so that initial modeling efforts can be focused on the functional aspects.

5.3.2.1 A General Delay Time Model

To model the delay time of a circuit, consider a generic well formed combinational circuit that performs an operation f on a number of input signals d_i and produces an output signal q as shown in Figure 5-5(a). The total delay time of the circuit is modeled as two parts as illustrated in Figure 5-5(b). The first part is called the Data_Valid time and is modeled using the VHDL inertia

(a) A generic well formed combinational circuit f

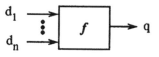

(b) Delay time model

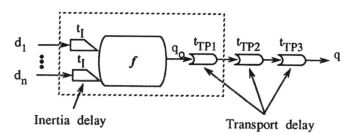

Inertia delay   Transport delay

(c) Timing diagram with t_I = 1ns

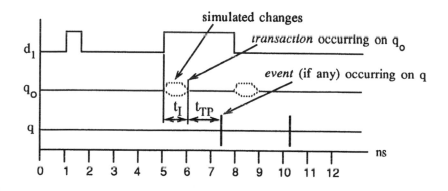

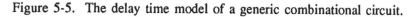

Figure 5-5. The delay time model of a generic combinational circuit.

delay model. This part characterizes the minimum stable duration t_I of the input signal before the operation f can be performed. The second part (t_{TP}) is the pure propagation delay and is modeled by three transport delay components. The first one (t_{TP1}) is the pure propagation delay inherent in the circuit. The other two components model the effects of external capacitive loadings due to the connecting input devices and interconnects.

The Data_Valid time part of this delay time model can be thought of as a low pass filter that eliminates glitches of duration shorter than the specified time t_I. As shown in the timing diagram of Figure 5-5(c), the first pulse of d_1 does not make it through the filter since its duration is less than t_I. Once through the filter, however, the input changes become valid and activate the process f. The result is the occurrence of a *transaction* on q_0, which may or may not cause a change of value in q_0. If a change does occur, then the change, called an *event*, will propagate to q after t_T. This chain of events is illustrated by the second pulse of d_1 in Figure 5-5(c). This example also illustrates the crucial role of the Data_Valid time parameter. In this work, this parameter is assumed to be constant. In reality, however, it varies dynamically depending on other factors cited previously. An aggressive design may exceed this limitation and result in either undetected (specified time too long) or non-existent (specified time too short) glitches on the output signal.

5.3.2.2 Parameter Values

The delay time model presented in the last section has four components. Essentially, the first two (the inertia delay and the inherent transport delay) may be derived from data books of cell libraries provided by ASIC vendors. In this work, if the setup time of a component is available from data books, then that time is used as the inertia delay parameter for that component. Otherwise, the inertia delay and the inherent transport delay are estimated from the delay time given in the data book with consideration of the actual circuit schematic. For simple logic gates, the delay time given in the data book is treated as the inherent transport delay in most cases.

The last two delay components depend on the physical design. For the third component, the transport delay due to fanout, the value is derived from the circuit after the detailed design is complete. The fourth component, the transport delay due to the interconnect capacitive loading, is estimated based on statistics following the rules suggested in [LSI87a]. With this approach, the interconnect delay is a function of both the circuit's size and the fanout of the signal. If the terminals of an interconnect all reside in the same module, then the total area of the module is first computed from the estimated gate count. The area of the module together with the fanout of the signal is then used to obtain an estimate of the wire's equivalent loading. If the terminals of a signal do not reside in the same module, then the area is computed as a function of the areas of the modules involved.

From the foregoing explanation, it should be obvious that the modeling and testing efforts for the majority of the circuit modules will be conducted in a situation where information of the delay time parameter values is incomplete. Nonetheless, redundant modeling effort can be minimized if the unknown part of the model is separated from the known part, and if freedom is allowed for these unknown values be specified in a progressive and seamless fashion. This requires a careful planning of the simulations, which in turn mandates the formulation of a programming strategy for specifying the delay time parameters.

5.3.2.3 Programming Strategy

Consider the situation of a circuit containing two modules as shown in Figure 5-6. The two modules may or may not be in the same hierarchical level.

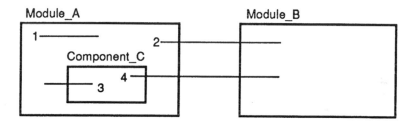

Figure 5-6. Four general signal connecting situations.

Inside Module_A, a component C has been instantiated. The four signals involved in Module_A illustrate various situations. Signal 1 is the simplest case where the delay time parameter values for this signal can be determined once the detailed design of Module_A is completed. The derived values are then directly specified in the signal assignment statement within the architecture body of Module_A's entity. Signal 2 originates in Module_A but the parameter values are partially dependent on Module_B. Both signals 3 and 4 originate in Component_C. While parameter values for signal 3 can be determined from Module_A alone, those for signal 4 cannot. The programming strategy described below deals with how the delay time information for the latter three signals can be efficiently managed using the VHDL construct configuration.

Following the bottom-up simulation approach, Component_C as a leaf-node will be first modeled and tested. The delay time parameters for signals 3 and 4 are specified in the generic interface list of Component_C's entity declaration as shown in Code 5-2(a). The entity declaration and architecture body of Module_A are shown in Code 5-2(b). The Module_A generic interface list contains only the delay time parameters for signal 2.

For Module_A to use Component_C, a component template Temp_C is first declared inside the declaration region of Module_A's architecture body. Note that while the port interface list of this declaration is exactly the same as

(a) Component_C   (b) Module_A

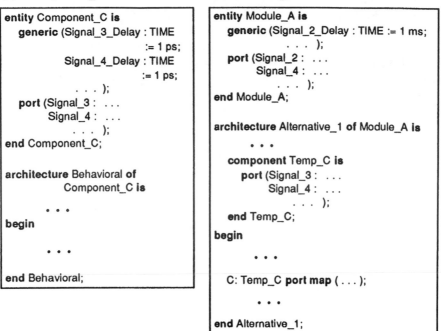

```
entity Component_C is
   generic (Signal_3_Delay : TIME
                        := 1 ps;
            Signal_4_Delay : TIME
                        := 1 ps;
            . . . );
   port (Signal_3 :  . . .
         Signal_4 :  . . .
            . . . );
end Component_C;


architecture Behavioral of
            Component_C is

       . . .
begin

       . . .

end Behavioral;
```

```
entity Module_A is
   generic (Signal_2_Delay : TIME := 1 ms;
            . . . );
   port (Signal_2 :  . . .
         Signal_4 :  . . .
            . . . );
end Module_A;


architecture Alternative_1 of Module_A is

   . . .

   component Temp_C is
      port (Signal_3 :  . . .
            Signal_4 :  . . .
               . . . );
   end Temp_C;
begin

       . . .

   C: Temp_C port map ( . . . );

       . . .

end Alternative_1;
```

Code 5-2. A sketch of VHDL descriptions of (a) Component_C and
(b) Module_A.

that of Component_C, no generic interface list is present. Thus, the use of
Temp_C in Module_A at this point is purely abstract with respect to delay
time. As Module_A is ready for initial testing, delay time information of
signals 2 and 4 will still be incomplete, but that of signal 3 can be derived.
This partial information of Component_C's delay time values is specified
through the configuration statement as shown in Code 5-3. In this
configuration, the specification of signal_3 delay time parameter is complete,
but that of signal_4 is only an assumed value. The test program selects this
configuration in its binding of the Module_A instance for simulation. When the
design of other modules is finished and information of other parameter values
becomes available, the old parameter values of the configuration Test_A can be
modified to form a new configuration as shown in Code 5-4. If Module_A is
used in other higher level circuit modules, then new simulation models can be
easily constructed by replacement.

The programming strategy just described not only saves programming and
computation efforts, it also provides a structure for investigating the effects of
using components of different performance characteristics and for simulation
with back annotation after the physical design is complete.

```
configuration Test_A of Module_A is
  for Alternative_1
    for C : Temp_C
      use WORK.Component_C (Behavioral)
        generic (Signal_3_Delay => 1.5 ns,    -- Derived from circuit
                 Signal_4_Delay => 2 ns,      -- An assumed value
                    . . . );
    end for;

      . . .

  end for;
end Test_A;
```

Code 5-3. Configuration for the testing of Module_A.

```
configuration The_Real_Thing_A of Module_A is
  for Alternative_1
    for C : Temp_C
      use WORK.Component_C (Behavioral)
        generic (Signal_3_Delay => 1.5 ns,    -- Derived from circuit
                 Signal_4_Delay => 3.3 ns,    -- Also a derived value
                    . . . );
    end for;

      . . .

  end for;
end The_Real_Thing_A;
```

Code 5-4. Configuration for the real use of Module_A.

5.4 VHDL Description of the IKS Chip

This section presents the VHDL description of the IKS chip. An overview of the circuit hierarchy and the organization of the libraries are given first. The circuit modules within each library are then described.

5.4.1 Overview

The circuit hierarchy of the IKS chip has five levels as shown in Figure 5-7. A library is created for each level below the root library IKS_Chip. Configurations of an entity are analyzed in the same library. Low-level building block circuits such as latches, multiplexers, counters, and shifters are put into the library **Macros**. They represent circuits that can be selected from commercial gate array libraries and thus are treated as leaf-nodes. One characteristic of modules at this level is that they will be instantiated by other circuit modules in all other levels. In contrast, for all higher levels, a circuit at

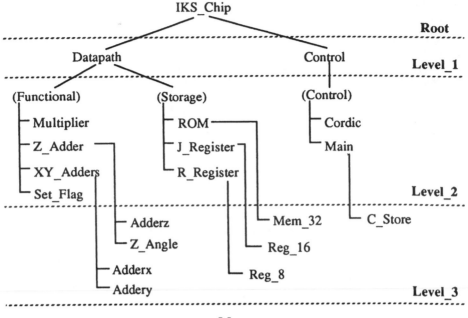

Macros

Figure 5-7. The circuit hierarchy of the IKS chip.

level i will be used only by another module at the immediate higher level i-1 as a result of the top-down design process. Since these building block circuits are simple and well understood, they are described first so that the focus can be placed on the basics of modeling hardware behavior with VHDL.

Then, rather than moving to the next level up (the **Level_3** library), the circuit modules in the **Level_2** library are described. However, the description may descend into relevant Level_3 library units directly from individual modules when the circuit decomposition calls for doing so. This not only allows the focal point to be switched back temporarily to the design aspect, it also serves to illustrate how the language constructs can be used to facilitate the design process in a top-down manner. The rest of the libraries will be described in a bottom-up fashion again until the root of the circuit hierarchy is reached.

In addition to the Custom_Types package, the root library also contains a package Cells, which includes component declarations for instantiation of design units in the library Macros. Level_1 and Level_2 libraries also contain a package Modules, which includes declarations of components corresponding to the design entities in their respective libraries. The component declarations in these packages follow the scheme outlined in Section 5.3.2.3.

Notation

A notation system for labeling circuit diagrams has evolved in the process of describing the circuits in VHDL programs. The system uses different symbolic styles to convey design information. In each circuit diagram, a box of dotted lines establishes the boundary of the module as a VHDL design entity. The box is labeled by a level number that identifies the module's hierarchical level and thus the library in which the corresponding design entity resides. Wires in the circuit diagrams become signals in the VHDL programs and each wire is given a signal name. The names inside the boxes indicate what the designated signals are called inside the entity. The names outside the box indicate what they are called at the next higher level when the entity is instantiated and connected as a component. If the names of a signal inside and outside the box (entity) are the same, then only one name will appear. In the circuit diagrams, a fine line indicates a single wire while a thicker line indicates an aggregate of wires. If that aggregate of wires is always interpreted collectively, for example as a number of a certain range, then the name is presented in normal typeface to denote a scalar. Otherwise, the signal is a **vector** and is denoted by **boldface**. If the signal is an element of a vector, then the element signal will be denoted, in normal typeface, by the same vector signal name with a subscript indicating its position in the vector. The *italic* typeface is used for signals whose functions are addresses or control-related. A control vector is thus denoted by *bold italic*. In addition, the style of filling a rectangle is used to indicate the controlling clock phase of a latch, with a stipple pattern and a diagonal pattern denoting Ø1 and Ø2, respectively.

5.4.2 The Macros Library

   Major circuit modules contained in the Macros library include latches, controlled input buffers, tristate output buffers, multiplexers, counters, shifters, and transmission gates. In addition, a two-phase non-overlapped clock and a number of macrocells, including two PLAs and two ROM cells for storing constants, are also contained in this library. The port interface list of each entity declaration that is to be used by more than one higher level module is duplicated in a corresponding component declaration statement and is collected in the package Cells. This package is analyzed (by executing the command of vhdl in Figure 5-3) into the library IKS_Chip.

5.4.2.1 Latches

   Three types of d-latches are used in the IKS chip. The first one has the scan path built-in, which can form a shift register during testing and is used in latching the control signals from the control store. The second type does not have the scan path built-in but rather has a reset control. This type is used in the main control module and also in register cells where the reset capability can facilitate memory testing. The third type is the simple latch without built-in scan path or reset capabilities.

(a) Latch_32

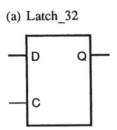

(b) Latch_r

Figure 5-8. Circuit
symbols of Latch_32
and Latch_r.

```
Library IKS_Chip; use IKS_Chip.Custom_Types.ALL;
entity Latch_32 is
   generic (Data_Valid : TIME := 1.1 ns;
              TP1 : TIME := 1.5 ns;
              TP2, TP3 : TIME := 0.5 ns;
              Gate_Delay : TIME := 0.5 ns);
   port (D : in BIT_32;
           C : in BIT;
           Q : out BIT_32 := 0);
end Latch_32;
architecture Behavioral of Latch_32 is
   signal Data : BIT_32 := 0;
   signal action : BIT;
   constant TP_Delay : TIME := TP1 + TP2 + TP3;
begin
   Data <= D after Data_Valid;
   action <= C after Gate_Delay;
   Act: block (action = '1')
            begin
                Q <= guarded transport Data after TP_Delay;
            end block;
end Behavioral;


entity Latch_r is
   generic ( . . .   Same as Latch_32 . . . );
   port (D, C, R_bar : in BIT;
           Q : out BIT);
end Latch_r;
architecture Behavioral of Latch_r is
   signal Data, action : BIT;
   constant TP_Delay : TIME := TP1 + TP2 + TP3;
begin
   Data <= D after Data_Valid;
   action <= C after Gate_Delay;
   Act: process (action, R_bar, Data)
            begin
                if R_bar = '0' then
                    Q <= transport '0' after TP_Delay;
                elsif action = '1' then
                    Q <= transport Data after TP_Delay;
                end if;
            end process;
end Behavioral;
```

Code 5-5. VHDL descriptions of Latch_32 and Latch_r.

The scan path is used only during testing for the first type of latch. During normal operation, they work essentially the same way as the third type except that they have different delay characteristics. Since the goal of the design at this stage is to model the circuit behavior of the normal operation, test-related capabilities are provided but not implemented. That is, the port interface list in the entity declaration of this type of latch still has the test clock, the scan-in, and the scan-out ports, but the architecture body is the same as the third type. The circuit symbols for the second and third types are shown in Figure 5-8 and their corresponding VHDL descriptions are listed in Code 5-5. The data type used by the latch model is indicated by the suffix in the entity name. The delay time parameters are specified in the generic interface list following the model explained in Section 5.3.2.1. That is, the Data_Valid time is derived from the set_up time and TP1, TP2, and TP3 account for the inherent transport delay and the transport delays due to the fanout and interconnect loadings, respectively.

5.4.2.2 Controlled Input Buffer

The controlled input buffer consists of a transmission gate connected in series with an input buffer. This circuit is used in various modules to receive data from the busses. It is assumed that when the transmission gate is turned off, the input node of the buffer will retain its charge at a valid logic level for a period no less than the clock cycle. Figure 5-9 shows the logic diagram and Code 5-6 lists the corresponding VHDL description.

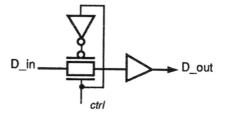

Figure 5-9. Logic diagram of a controlled input buffer.

```
library IKS_Chip; use IKS_Chip.Custom_Types.ALL;
entity In_Buffer_32 is
    generic (Data_Valid : TIME := 0.5 ns;
                 TP1 : TIME := 0.9 ns;
                 TP2, TP3 : TIME := 0.5 ns;
                 Discharge : TIME := 50 us);
    port (D_In : in BIT_32;
            D_out : out BIT_32 := 0;
            ctrl : in BIT);
end In_Buffer_32;

architecture Behavioral of In_Buffer_32 is
    signal Data : BIT_32 := 0;
    constant TP_Delay : TIME := TP1 + TP2 + TP3;
begin
    Act: block (ctrl = '1')
            begin
                Data <= guarded D_In after Data_Valid;
            end block;
    D_out <= transport Data after TP_Delay;
    assert not (ctrl = '0' and ctrl'STABLE(Discharge))
        report "Data may not be valid"
        severity WARNING;
end Behavioral;
```

Code 5-6. VHDL description of the controlled input buffer.

5.4.2.3 Output Buffer

The logic diagram and the VHDL description of the output buffer are shown in Figure 5-10 and Code 5-7, respectively. The circuit has incorporated a clocked control signal. Associated with the output signal (not shown in the circuit diagram) is a virtual_signal state, which will be set to 0 to indicate the open connection state whenever the AND gate output drops. The actual number of bits of the data signal is 32. To reduce the loading on the AND gate output, a two-level buffering scheme (1-NAND-to-4-INV) is used. The delay time parameters for the AND gate has already taken this into account.

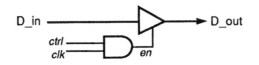

Figure 5-10. Logic diagram of an output buffer.

```
library IKS_Chip; use IKS_Chip.Custom_Types.ALL;
entity Tristate_Out_32 is
  generic (AND_Gate_Delay : TIME := 1.6 ns;
                Data_Valid : TIME := 0.2 ns;
                TP1 : TIME := 0.5 ns;
                TP2, TP3 : TIME := 0.5 ns);
    port (D_in : in BIT_32;
           D_out : out BIT_32 := 0;
           state : out VIRTUAL_SIGNAL := 0;
           ctrl, clk_in : in BIT);
end Tristate_Out_32;

architecture Behavioral of Tristate_Out_32 is
    signal en : BIT;
    signal Data, src : BIT_32 := 0;
    constant TP_Delay : TIME := TP1 + TP2 + TP3;
begin
    en <= transport ctrl and clk_in after AND_Gate_Delay;
    Data <= D_in after Data_Valid;
    src <= transport Data after TP_Delay;
    process (src, en)
    begin
      if en = '1' then
          D_out <= src,
          state <= 1;
      else
          state <= 0;
      end if;
    end process;
end Behavioral;
```

Code 5-7.   VHDL description of the controlled output buffer.

5.4.2.4 Multiplexer

The circuit symbol and VHDL description of a 2-to-1 multiplexer are shown in Figure 5-11 and Code 5-8, respectively. In the VHDL architecture body, a process is used to convert the value of the *sel* signal (of type BIT) into the type SOURCES_2 so that the conditional select statement (M2) can be applied to make it more expressive. The 4-to-1 multiplexers are constructed similarly.

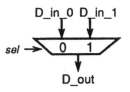

Figure 5-11. Circuit symbol of a 2-to-1 multiplexer.

```
library IKS_Chip; use IKS_Chip.Custom_Types.ALL;
entity Mux_2_to_1 is
   generic (Data_Valid : TIME := 0.7 ns;
            TG_Delay : TIME := 0.3 ns;
            TP1 : TIME := 0.3 ns;
            TP2, TP3 : TIME := 0.5 ns);
   port (D_in_0 : in BIT;
         D_in_1 : in BIT;
         D_out : out BIT;
         sel : in BIT);
end Mux_2_to_1;

architecture Behavioral of Mux_2_to_1 is
   signal src : SOURCES_2;
   signal D0, D1 : BIT;
   constant TP_Delay : TIME := TP1 + TP2 + TP3;
begin
   Act: process (sel)
         begin
            if sel = '0' then
               src <= src0 after TG_Delay;
            else
               src <= src1 after TG_Delay;
            end if;
         end process;
   D0 <= D_in_0 after Data_Valid;
   D1 <= D_in_1 after Data_Valid;
   M2: with src select
         D_out <= transport D0 after TP_Delay when src0,
                            D1 after TP_Delay when src1;
end Behavioral;
```

Code 5-8. VHDL description of the 2-to-1 multiplexer.

5.4.2.5 Counter

Counter circuits are usually available from gate array libraries. The circuit symbol and the VHDL description of a counter are shown in Figure 5-12 and Code 5-9, respectively. The counter delay is extracted from the data sheet and

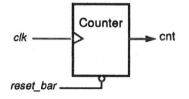

Figure 5-12. Circuit symbol of a counter circuit.

```vhdl
library IKS_Chip; use IKS_Chip.Custom_Types.ALL;
entity Counter is
   generic (Num_Bit : INTEGER := 5;
            CNT_Delay : TIME := 4 ns;
            Reset_Delay : TIME := 1 ns;
            TP1 : TIME := 1 ns;
            TP2, TP3 : TIME := 0.5 ns);
   port (reset_bar, clk : in BIT;
         cnt : buffer NATURAL := 0);
end Counter;

architecture Behavioral of Counter is
   signal new_cnt : NATURAL := 0;
   constant MAX_CNT : NATURAL := 2 ** Num_Bit;
   constant TP_Delay : TIME := TP1 + TP2 + TP3;
begin
   cnt <= transport new_cnt after TP_Delay;
   Count: process
           begin
             while reset_bar = '1' loop
               wait on reset_bar, clk;
               if (reset_bar= '0') then
                  new_cnt <= 0 after Reset_Delay;
               elsif (clk = '1') then
                  new_cnt <= (cnt + 1) mod MAX_CNT after CNT_Delay;
               end if;
             end loop;
             wait until reset_bar = '1';
           end process Count;
end Behavioral;
```

Code 5-9. VHDL description of the counter circuit.

specified in accordance with the delay time model discussed previously. Note that the interface allows the number of bits to be specified during component instantiations.

5.4.2.6 Shifter

Barrel shifter circuits are also available from gate array libraries. In general, right-shift and left-shift use the same circuit but the ports for data input and output are reversed. Since the IKS chip uses the right-shift operation only, no extra circuits are needed to direct the I/O. The circuit symbol and the VHDL description are shown in Figure 5-13 and Code 5-10, respectively. In the

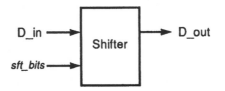

Figure 5-13. Circuit symbol of a shifter circuit.

```
library IKS_Chip; use IKS_Chip.Custom_Types.ALL;

entity Barrel_Shifter is
   generic (Decode : TIME := 2 ns;
             Shift_Delay : TIME := 4 ns;
             TP1 : TIME := 0.5ns;
             TP2, TP3 : TIME := 1 ns);
   port (D_in : in BIT_32;
          D_out : out BIT_32 := 0;
          sft_bits : in BIT_5);
end Barrel_Shifter;

architecture Behavioral of Barrel_Shifter is
   signal Data : BIT_32 := 0;
   signal num : BIT_5 := 0;
   constant TP_Delay : TIME := TP1 + TP2 + TP3;
begin
   num <= sft_bits after Decode;
   Shift: process (D_in, num)
           variable value : INTEGER;
         begin
           value := D_in / 2 ** num;
           Data <= value after SHIFT_Delay;
         end process;
   D_out <= transport Data after TP_Delay;
end Behavioral;
```

Code 5-10. VHDL description of the shifter circuit.

architecture body, the right-shift operation is modeled by division, since the operands of the shifter in the IKS chip are always integers.

5.4.2.7 Transmission Gate

VHDL allows description of hardware behavior down to the gate level, but modeling of transmission gates (TGs), which operate at the more primitive transistor level, is not supported [Sha86, LRM88]. To overcome this limitation, a technique for modeling the TG's behavior in VHDL for bus-oriented architecture designs has been developed [Leu89b]. The technique is based on the concept of a *virtual_signal* to encode the nature of the connections to the TG so that the TG can configure the dataflow direction dynamically.

Table 5-2. Virtual_Signal Values and Connection Types.

Virtual_Signal Value	Connection Type (as seen by the receiver)
−1	reader
0	open
1	driver
2	request status

The meanings of the virtual_signal values are tabulated in Table 5-2. A virtual_signal value of 1 designates an information source (driver) and a value of -1 designates a load (reader). A value of 0 indicates physically an open connection, electrically a high-impedance state, and logically a "don't care" condition. The virtual_signal value 2 does not specify a connection type. It is used by the TGs as an interrogation signal to configure the data flow dynamically. In addition, the circuit devices to which the TGs are connected are classified either as primary devices or secondary devices. A primary device can originate a virtual_signal value that specifies a connection type while a secondary device can only pass along the virtual_signal value. Five basic primary devices and the ranges of their virtual_signal values are listed in Table 5-3.

The VHDL TG model is described in terms of data structure design and semantics design. The data structure reflects the two abstract properties of the transmission gate I/O ports, namely, the information about the data type and the nature of the connection. The semantics describe operations on the data structure. The interpretation of the semantics constitutes the dynamic behavior of the transmission gate.

Without loss of generality, the TG circuit is considered as a three-port device with data I/O ports, Port_0 and Port_1, controlled by a single signal, *ctrl*. The TG's circuit symbol and the graphic representation of its data structure are shown in Figure 5-14. The control port *ctrl* is of type BIT and each I/O port

Table 5-3. Primary Devices and their Ranges of Virtual_Signal Values.

Primary Device	Virtual_Signal Value Range
gate input	$\{-1\}$
gate output	$\{1\}$
tristate output	$\{0, 1\}$
bi-directional buffer	$\{-1, 1\}$
bi-directional buffer with tristate output	$\{-1, 0, 1\}$

in the circuit symbol is translated into four signal ports accessible from the outside world. Among them, the port Data_i of an appropriate type represents the "official" value of Port_i. The buffer mode is used for the port Data_i so that it can be read from both inside and outside. An auxiliary input port and a resolution process are created to determine the value. Since the language has no restriction on reading a port value from outside an entity, the accomplishment of "official" status of Data_i can only rely on programming disciplines. In other words, it is the programmer's responsibility to refrain from reading Di_in anywhere in the program. In addition to the two data ports, each I/O port also

(a) Circuit symbol

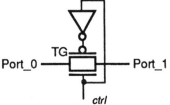

(b) Graphic representation of the data structure

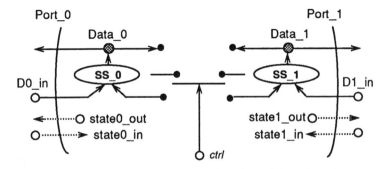

Figure 5-14. Circuit symbol and data structure of the TG model.

has two ports of type VIRTUAL_SIGNAL, as indicated by the dotted lines. These signals do not correspond to physical wires. They are conceptual devices for encoding the connection types so that the direction of data flow can be specified in the semantics of the TG body.

The semantics of the TG behavior are partitioned into three parts of state transition, steady states, and connecting protocol. The first two are realized by the process statements within the TG architecture body. The third part stipulates how the outside world interacts dynamically with the TG and is implemented by both the TG and the connecting processes.

Code 5-11 shows the TG model's VHDL entity declaration and the organization of its architecture. The architecture statement part contains three processes statements. They implement the TG's state transition and the steady states for Port_0 and Port_1.

```
entity Transmission_Gate is
  port (Data_0, Data_1 : buffer INTEGER := 0;
        D0_in, D1_in : in INTEGER;
        state0_out, state1_out : out VIRTUAL_SIGNAL := 0;
        state0_in, state1_in : in VIRTUAL_SIGNAL;
        ctrl : in BIT);
end Transmission_Gate;

architecture Behavioral of Transmission_Gate is
begin
  ST: process
        • • •    (Code implementing State_Transition)
      end process ST;
  SS_0: process
           • • •   (Code implementing Steady_States for Port_0)
         end process SS_0;
  SS_1: process
           • • •   (Code implementing Steady_States for Port_1)
         end process SS_1;
end Behavioral;
```

Code 5-11. VHDL description of the transmission gate.

The state of a TG instance is defined by the signal *ctrl* and the four virtual_signals. Since changes in data signals ultimately originate from the external world, they are not considered as constituents of the TG's internal state. The process statement ST (Code 5-12) implements the state transition diagram of Figure 5-15.

Initially, the TG is in a state of open connection characterized by a *ctrl* signal value of '0' and all four virtual_signals equal to 0. This is also the state to which the TG will return from other non-error states whenever the signal

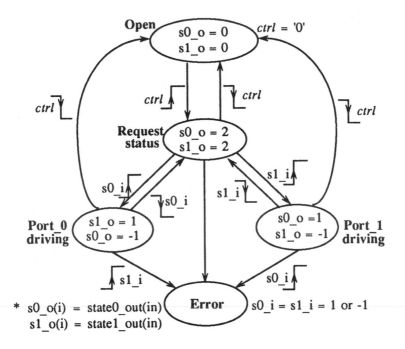

Figure 5-15. The state transition diagram of the TG.

ctrl experiences a '1' to '0' transition. When the signal *ctrl* rises from '0' to '1', the TG will set the output virtual_signals of both I/O ports to 2 and then monitor the responses on the two input virtual_signals. A single driving signal (virtual_signal value 1) from either port will cause the TG to enter one of the two connecting (driving) states. At any time, the two input virtual_signals having the same value of 1 (both driving) or -1 (both reading) will cause the TG to enter the error state and the simulation will terminate. Essentially, the switching process performs constraint violation checks and executes the protocol on the part of the TG to configure the data flow direction.

Among the four allowed states in the state transition diagram, only three correspond to physical configurations. They are the open state and the two connecting states — one from Port_1 to Port_0 and the other from Port_0 to Port_1. The implementation of steady states is manifested as the problem of assigning the right value at the right time to the I/O port's "official" value holder. This is done by the process statement SS_0 and SS_1 for Port_0 and Port_1, respectively. The process statement that implements the steady states for Port_0 is listed in Code 5-13.

The simulated behavior of the TG inherently depends on the nature of the connection. The protocol defines how this information is passed to the TG during simulation. The protocol involves both the TG and the connecting device, with the connecting device playing a passive role. The connecting device also has

```
ST: process
       variable s0, s1 : VIRTUAL_SIGNAL;
    begin
       state0_out <= 0;                -- TG in open state
       state1_out <= 0;
       wait until ctrl = '1';
       while ctrl = '1' loop
          state0_out <= 2;             -- Request status
          state1_out <= 2;
          wait on state0_in, state1_in, ctrl;
          if ctrl = '1' then
             s0 := state0_in;
             s1 := state1_in;
             assert not ((s0 = 1) and (s1 = 1))
                report "Both ends of the TG are driving"
                severity ERROR;
             assert not ((s0 = -1) and (s1 = -1))
                report "Both ends of the TG are reading"
                severity ERROR;
             if s1 = 1 then             -- Port 1 driving
                state0_out <= 1;
                state1_out <= -1;
                while ctrl = '1' loop
                   wait on state0_in, state1_in, ctrl;
                   assert state0_in /= 1
                      report "Both ends of the TG are driving"
                      severity ERROR;
                   if state1_in /= 1 then exit; end if;
                end loop;
             elsif s0 = 1 then          -- Port 0 driving
                state0_out <= -1;
                state1_out <= 1;
                while ctrl = '1' loop
                   wait on state0_in, state1_in, ctrl;
                   assert state1_in /= 1
                      report "Both ends of the TG are driving"
                      severity ERROR;
                   if state0_in /= 1 then exit; end if;
                end loop;
             end if;
          end if;
       end loop;
    end process ST;
```

Code 5-12. The process statement implementing the state transition.

processes to handle the state transitions and the steady states separately. The protocol, implemented by these two processes, can be described algorithmically as shown in Figure 5-16.

The TG model has been tested as a stand-alone device with integer data type for the I/O ports, which are connected to separate controlled driving sources. The TG's control port is connected to a clock signal, which turns on for 30 ns every 100 ns starting at 50 ns. Data are then driven to the two ports with different timing characteristics to test various dataflow cases. The simulation results have confirmed that the model exhibits the intended behavior. To further

```
SS_0: process
        begin
            if ctrl = '0' then              -- TG OFF
                wait on ctrl, D0_in;
                if not D0_in'STABLE then
                    Data_0 <= D0_in;
                end if;
            else                            -- TG ON
                if state1_in = 1 then       -- Port 1 driving
                    Data_0 <= Data_1;
                    wait on Data_1, state1_in, ctrl;
                else
                    wait on D0_in, state1_in, ctrl;
                    if not D0_in'STABLE then
                        Data_0 <= D0_in;
                    end if;
                end if;
            end if;
        end process SS_0;
```

Code 5-13. The process statement implementing the steady states for Port_0.

test the robustness of the TG model, two TGs are instantiated and connected in series. The TG-to-TG connections are made directly from one TG's out-mode port to the other's in-mode port. The control ports of both TGs are tied to the same clock signal and the same input pattern of the one-TG experiment is applied to the open ends of the two-TG circuit. As expected, the simulation result is the same as before except that some of the signals settle down at a greater simulation (delta) cycle.

- State_Transition process:
  { if virtual_signal state_out (from TG) changes to 2 then
      report connection_type to virtual_signal state_in;
  else if value of state_out drops from 1 then
      send the current state_out value to the readers;
  end if; };
- Steady_States process:
  { if virtual_signal state_out (from TG) = 1 then
      assign the current Data value of the TG to readers if any;
  else
      assign the current driver's value to D_in if a driver exists;
  end if; }.

Figure 5-16. Algorithmic description of the connecting
processes executing the TG's protocol.

5.4.3 The Level_2 Library

Circuit modules contained in this library can be divided into three categories of functional units, storage, and controls. The complexities of some circuit modules make further decomposition necessary. The presentation of these modules follows the logic structure of this decomposition. Functional unit circuits are described first, followed by storage modules and control units.

5.4.3.1 The Multiplier Module

Figure 5-17 shows the circuit diagram of the Multiplier. Signal names shown inside the Level_2 box are declared as **port** names in the VHDL interface. Names outside the box are the actual signals (declared at a higher level) connecting to these ports. The multiplier core consists of a two-stage pipeline implementation (see Section 4.3.2.2.4). It accepts two 24-bit integers and returns a 32-bit product, all in two's complement representation and conforming to the data formats as specified in Figure 4-9.

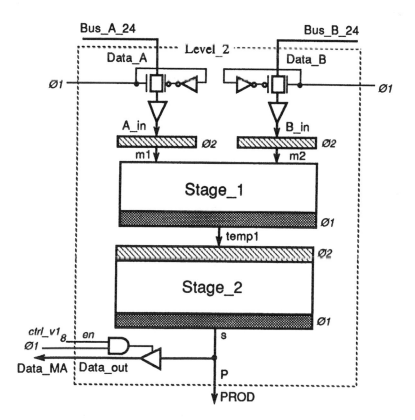

Figure 5-17. Circuit diagram of the Multiplier.

The VHDL description of the Multiplier is listed in Code 5-14. The first two statements, called context items, make ALL the declarative items within the packages Custom_Types and Cells, both in the library IKS_Chip, visible to the entity Multiplier. (These two statements are present in all other high-level design entities but will be omitted in the program listings that follow.) Multiplication is simulated by the function mpy, which is declared and specified in the declarative part of the architecture body. Checking mechanisms have been incorporated into the code to detect for the overflow conditions. The reading inside the entity makes it necessary to use the **buffer** mode for port P. Since all the output ports of the latches are declared to be **out** mode, a local signal S is used to pass the value of the output signal s from latch L1_s to P.

```
library IKS_Chip; use IKS_Chip.Custom_Types.ALL, IKS_Chip.Cells.ALL;
entity Multiplier is
  generic (MPY_1_Delay : TIME := 40 ns;
           MPY_2_Delay : TIME := 25 ns);
    port (Data_A, Data_B : in BIT_24;
          P : out BIT_32 := 0;
          Data_out : out BIT_32;
          D_state : out VIRTUAL_SIGNAL;
          en, phase_1, phase_2 : in BIT);
end Multiplier;

architecture Dataflow of Multiplier is
  function mpy (signal m1, m2 : in BIT_24) return BIT_32;
  function mpy (signal m1, m2 : in BIT_24) return BIT_32 is
    constant factor : INTEGER := 2**14;
    constant bound : INTEGER := 2**17;
    variable x1, x2, x1_u, x1_l, x2_u, x2_l, sign_m1, sign_m2, p, p1 : INTEGER;
  begin

    . . . (Code that implements the multiplication of two 24-bit integer numbers)

  end mpy;
  signal A_in, B_in, m1, m2 : BIT_24;
  signal temp, temp1, temp2, result, s : BIT_32 := 0;

begin
  In_A: Inbuf_24 port map (Data_A, A_in, phase_1);
  In_B: Inbuf_24 port map (Data_B, B_in, phase_1);
  L2_m1: Latch_24 port map (A_in, phase_2, m1);
  L2_m2: Latch_24 port map (B_in, phase_2, m2);
  Stage_1: temp <= mpy (m1, m2) after MPY_1_Delay;
  L1_temp1: Latch_32 port map (temp, phase_1, temp1);
  L2_temp2: Latch_32 port map (temp1, phase_2, temp2);
  Stage_2: result <= temp2 after MPY_2_Delay;
  L1_s: Latch_32 port map (result, phase_1, s);
  P <= s;
  T_P: Tristate_32 port map (s, Data_out, D_state, en, phase_1);
end Dataflow;
```

Code 5-14. VHDL description of the Multiplier.

The count of effective gates for this module is 5288. With an estimated utilization of 80%, the practical gate count is 6610.

5.4.3.2 The Z_Adder Module

The Z_Adder has a dual role as a part of the MAC and also as a part of the cordic core. The MAC part is easy to implement with the provision of a direct link from the output of the multiplier. Hardware support for execution of the cordic algorithm presents some additional requirements on this module. First, the operation (+/−) is determined dynamically depending on the signs of the intermediate results during an iteration. This requirement can be satisfied by simply multiplexing the operation control signals from the control store (for normal operation) and from the cordic control unit. The second requirement involves some special processing of the operands during the cordic calculation.

To assure convergence of the cordic algorithm, the angle operands need to be preprocessed so that the resultant angles lie in either the first or the fourth quadrant. Furthermore, if the result is an angle, then post processing may be required to map the result back to the correct quadrant. Due to the fractional turn representation of the angles, the mapping of the angles between quadrants can be easily implemented by processing the two most significant bits only (see Figure 4-9). Since this can be done with only a few gates while the data is being transferred the costs of area and delay time are insignificant. Moreover, it can save a number of cycles and additional storage space.

The processing just described is put into a "black box" called Z_Angle with two control signals *zang* and *f* for specifying the required processing. The dataflow table is then modified with the notation of ang(*zang, f*) indicating the operations needed. From the dataflow table, operations required by the Z_Adder module are summarized in Table 5-4.

As shown in the table, the control signal Z_0 is used to select whether the operation control is taken from the control store (Z_1) or from the cordic control unit (A_z). Two other control signals *z1* and *z2* are used to select the operands. Note that when *z2* is equal to 2 ("10"), the addend is taken from the output of the Z_Angle block.

The circuit diagram and the corresponding VHDL description of the Z_Adder is shown in Figure 5-18 and Code 5-15, respectively. The black box Z_Angle, whose requirement has just been described, forms a Level_3 entity in the VHDL description. To simplify the description and also facilitate the reuse of designs, the 32-bit adder, together with its input and output latches, is taken out to form another Level_3 entity as indicated by the nested dotted box. Unlike other entities in the Macros library, these two entities have not been created. Thus, they are declared as components locally. The connections to these two components are specified through the instantiation statements in the statement part of the architecture body. The instantiation in these two cases are examples of the top-down design decomposition process supported by VHDL.

The count of effective gates for this module is 2809. With a utilization of 80%, the practical gate count is estimated to be 3511.

Table 5-4. Functional Table of the Z_Adder.

$z2$	$z1$	Z_0	SUM_Z = ($z1$) operation ($z2$)*
0	0	X	0
0	1	X	PROD
0	2	X	Bus_A
1	0	0/1	(Z_1 / A_z) Bus_B
1	1	0/1	PROD (Z_1 / A_z) Bus_B
1	2	0/1	Bus_A (Z_1 / A_z) Bus_B
2	0	0/1	(Z_1 / A_z) ang($zang$, f)
2	1	0/1	PROD (Z_1 / A_z) ang($zang$, f)
2	2	0/1	Bus_A (Z_1 / A_z) ang($zang$, f)

* $Z_0 = 0$ (1) $\Rightarrow$ operation = Z_1 (A_z);
if the value of Z_1 or A_z is 0 (1), the operation is + (−).

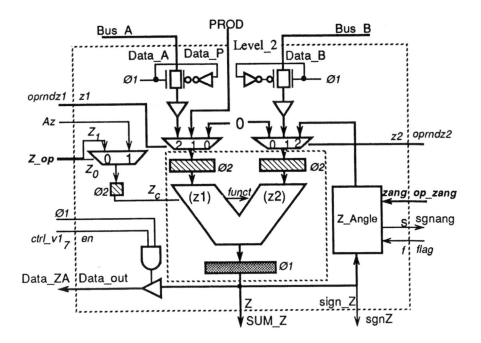

Figure 5-18. Circuit diagram of the Z_Adder.

```
entity Z_Adder is
  generic (Sign_TP_Delay : TIME := 1 ns);
  port (Data_A, Data_B, Data_P : in BIT_32;
        Z : out BIT_32 := 0;
        Data_out : out BIT_32;
        D_state : out VIRTUAL_SIGNAL;
        sign_Z, s : out BIT;
        z1, z2, Z_op, zang : in VECT_2;
        f, Az, en : in BIT;
        phase_1, phase_2 : in BIT);
end Z_Adder;

architecture Structural of Z_Adder is
  component Addor
    port (addend_1, addend_2 : in BIT_32;
          sum : out BIT_32;
          funct : in BIT;
          phase_1, phase_2 : in BIT);
  end component;
  component Angle
    port (D_in : in BIT_32;
          D_out : out BIT_32;
          z_ang_0, z_ang_1, f, phase_1 : in BIT;
          sign_zang : out BIT);
  end component;
  signal A_in, B_in, zsum, z1_in, z2_in, Zang_out : BIT_32;
  signal GND, open_src : BIT_32 := 0;
  signal Zc, Zc_1 : BIT;
begin
  In_A: Inbuf_32 port map (Data_A, A_in, phase_1);
  In_B: Inbuf_32 port map (Data_B, B_in, phase_1);
  M4_z1: Mux32_4 port map (GND, Data_P, A_in, open_src, z1_in, z1);
  M4_z2: Mux32_4 port map (GND, B_in, Zang_out, open_src, z2_in, z2);
  Add_Z: Adder port map (z1_in, z2_in, zsum, Zc, phase_1, phase_2);
  Z <= zsum;
  M2_Zc: Mux_2 port map (Z_op(0), Az, Zc_1, Z_op(1));
  L2_Zc: Latch port map (Zc_1, phase_2, Zc);
  T_Z: Tristate_32 port map (zsum, Data_out, D_state, en, phase_1);
  Ang_Blk: Angle port map (zsum, Zang_out, zang(1), zang(0), f, phase_1, s);
  Sgn_of_Z: process (zsum)
            begin
              if zsum < 0 then
                sign_Z <= transport '1' after Sign_TP_Delay;
              else
                sign_Z <= transport '0' after Sign_TP_Delay;
              end if;
            end process Sgn_of_Z;
end Structural;
```

Code 5-15. VHDL description of the Z_Adder.

The Adder Sub-module

The Adder sub-module has two input data links and one output data link. During phase-one, the input operands are latched and the result of the last computation is also available for output. The operation is controlled by the signal *funct*, which changes during the beginning of the phase-two clock and the signal should remain stable until the phase-two clock of the next cycle.

```vhdl
entity Adder_32 is
  generic (Adder_Delay : TIME := 25 ns);
  port (addend_1, addend_2 : in BIT_32;
        sum : out BIT_32;
        funct : in BIT;
        phase_1, phase_2 : in BIT);
end Adder_32;

architecture Functional of Adder_32 is
  signal a1, a2 : BIT_32;
  signal s : BIT_32 := 0;
begin
  L2_a1: Latch_32 port map (addend_1, phase_2, a1);
  L2_a2: Latch_32 port map (addend_2, phase_2, a2);
  Add: process (a1, a2, funct)
          variable x, y, z : INTEGER;
          constant bound : INTEGER := INTEGER'RIGHT/2;
        begin
          x := a1/2;
          y := a2/2;
          if funct = '0' then
            z := x + y;
          else
            z := x - y;
          end if;
          if z < 0 then z := -z; end if;
          assert z < bound
            report "Adder overflow"
            severity WARNING;       -- Don't want to stop the simulator
          if z < bound then
            if funct = '0' then
              s <= a1 + a2 after Adder_Delay;
            else
              s <= a1 - a2 after Adder_Delay;
            end if;
          else          -- Overflow
            s <= z*2 after Adder_Delay;
          end if;
        end process Add;
  L1_sum: Latch_32 port map (s, phase_1, sum);
end Functional;
```

Code 5-16. VHDL description of a 32-bit Adder.

The VHDL description of the 32-bit Adder circuit is listed in Code 5-16. The addition is implemented by a process statement with overflow checks performed during the simulation. This module is analyzed into the Level_3 library. In addition to the Z_Adder module, the XY_Adders will also instantiate this entity as components. Different delay time values for these modules are specified through configurations. The one used in the Z_Adder is configured as Adderz and the configuration statement is listed in Code 5-17. The parameter values are derived after the overall design is completed. This configuration is analyzed into the same Level_3 library and is the library unit to which the Adder instance (Add_Z within the Z_Adder) is bound when the Z_Adder itself needs to be configured for use in the architecture of Datapath at the next higher level.

The practical gate count of the Adder sub-module is estimated to be 2081.

```
library IKS_Chip, Macros; use IKS_Chip.Cells.All, Macros.ALL;
configuration Adderz of WORK.Adder_32 is
   for Functional
      for L2_a1 : Latch_32
         use entity Macros.Latch_32 (Behavioral)
         generic map (Data_Valid => 1.1 ns,
                      TP1 => 1.5 ns,
                      TP2 => 0.6 ns,
                      TP3 => 0.5 ns,
                      Gate_Delay => 0.6 ns);
      end for;
      for L2_a2 : Latch_32
         use entity Macros.Latch_32 (Behavioral)
         generic map (Data_Valid => 1.1 ns,
                      TP1 => 1.5 ns,
                      TP2 => 1.2 ns,
                      TP3 => 0.5 ns,
                      Gate_Delay => 0.7 ns);
      end for;
      for L1_sum : Latch_32
         use entity Macros.Latch_32 (Behavioral)  -- use high drive
         generic map (Data_Valid => 0.8 ns,
                      TP1 => 0.7 ns,
                      TP2 => 1.5 ns,
                      TP3 => 1.1 ns,
                      Gate_Delay => 0.9 ns);
      end for;
   end for;
end Adderz;
```

Code 5-17. VHDL description of the Adderz configuration.

The Z_Angle Sub-module

The Z_Angle sub-module performs three operations. The first is a null operation (no-op) in which the output remains unmodified. Both the second and the
third operations involve complementing the sign bit. The second one depends on
a *flag* signal, while the third one depends on the exclusive-or of the two most
significant bits of the input. These operation requirements are specified in Table
5-5. The circuit implementation of the functional table is shown in Figure 5-
19.

Table 5-5. Functional Table of the Z_Angle.

a_1	a_0	f	Data_out
X	0	X	Data_in
0	1	0	Data_in
0	1	1	complement sign bit
1	1	X	if $Z_{n-1} \neq Z_{n-2}$ then complement sign bit

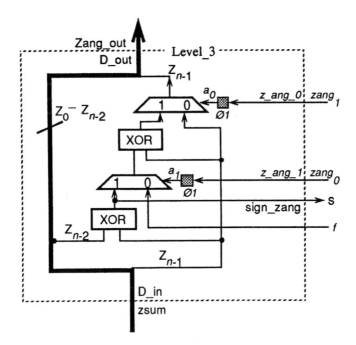

Figure. 5-19. Circuit diagram of the Z_Angle.

```vhdl
architecture Logic of Z_Angle is
   constant msb2_pos : INTEGER := INTEGER'RIGHT/2 + 1;
   constant msb2_neg : INTEGER := INTEGER'LEFT/2 - 1;
   constant half_turn : INTEGER := INTEGER'RIGHT + 1;
   signal z_n1, z_n2, sgn, a0, a1, s1, s2, s3 : BIT;
begin
  Extr_Sgn: process (D_in)        -- Extract the most significant two bits
            begin
                if D_in < 0 then
                    z_n1 <= '1';
                    if D_in > msb2_neg then
                      z_n2 <= '1';
                    else
                      z_n2 <= '0';
                    end if;
                else
                    z_n1 <= '0';
                    if D_in > msb2_pos then
                      z_n2 <= '1';
                    else
                      z_n2 <= '0';
                    end if;
                end if;
            end process Extr_Sgn;
  sign_zang <= sgn;
  sgn <= transport z_n1 xor z_n2 after XOR_Gate_Delay;
  M2_s1: Mux_2 port map (f, sgn, s1, a1);
  L1_a1: Latch port map (z_ang_1, phase_1, a1);
  L1_a0: Latch port map (z_ang_0, phase_1, a0);
  s2 <= transport z_n1 xor s1 after XOR_Gate_Delay;
  M2_s3: Mux_2 port map (z_n1, s2, s3, a0);
  Intrprt: process (s3, D_in, a0)
            begin
                if (a0 = '0') then  -- No processing required
                    D_out <= D_in;
                else                -- Look at s3 and sign of D_in
                  if s3 = '0' then     -- Output should be +ve, (I or II qrad)
                    if D_in < 0 then  -- But input is -ve (III or IV quad.) , so
                      D_out <= D_in + half_turn;
                    else              -- OK
                      D_out <= D_in;
                    end if;
                  else              -- Output should be -ve (III or IV quad.),
                    if D_in < 0 then   -- OK
                      D_out <= D_in;
                    else              -- Now input is +ve (I or II quad.), so
                      D_out <= D_in - half_turn;
                    end if;
                  end if;
                end if;
            end process Intrprt;
end Logic;
```

Code 5-18. VHDL architecture of the Z_Angle.

The VHDL architecture body of the Z_Angle module is listed in Code 5-18. There are two process statements in the statement part. The first one responds to the D_in signal and extracts the information of the two most significant bits. The second process interprets the results of the circuit's logic manipulation into the higher abstraction of a 32-bit integer value.

The Z_Angle sub-module contains only 28 actual gates.

5.4.3.3 The XY_Adders Module

The primary role of the XY_Adders is to perform the shift-and-add operation during the cordic iterations. For some of the IKS inputs whose corresponding positions are either out-of-range or near a singularity point, the mode 1 cordic operation may result in an operation attempting to take the square root of a negative number. This condition can be detected by checking the sign bit of the difference of the two operands (computed via the Z adder). If this condition occurs, a flag is set and the result is set to zero at the end of the square root operation. Due to its extensive use, a zero operand is hardwired and made available to the input latches via multiplexers. This reduces the demand on resources for data transfer. The functions that the X and Y adders perform are derived from the dataflow table and are summarized in Tables 5-6 and 5-7, respectively. The operation controls of the two adders (X_c and Y_c) are derived from the same control vector XY as specified in Table 5-8. Note that when $y2$ is equal to 1, the result of the Y adder output is equivalent to the result after a branch instruction.

The circuit diagram of the XY_Adders is shown in Figure 5-20. The sign bits of the second operand and the result of the Y adder are hardwired for connections outside the module. The sign information is needed in the preprocessing of the arctan operation and the calculations during the cordic iterations. Testability considerations have led to the addition of three multiplexers which enable the two adder outputs and the inputs of the two shifters to be accessed via the two busses during testing. In normal operation, the control signals of these three multiplexers are set to 0.

Table 5-6. Functional Table of the X Adder.

$x1$	$x2$	SUM_X
0	0	$(X_c)*2^{-i}*Y$
0	1	(X_c) Bus_B
1	0	SUM_X (X_c) $2^{-i}*Y$
1	1	SUM_X (X_c) Bus_B

* For the definition of X_c see Table 5-8.

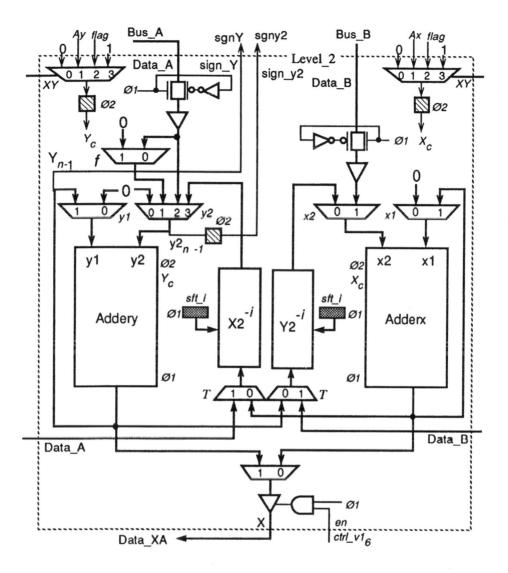

Figure 5-20. Circuit diagram of the XY_Adders.

The VHDL description of the XY_Adders is listed in Code 5-19. Similar to the Z_Adder module, an Adder template is declared locally in the architecture body. Two instances of the Adder, the Add_X and the Add_Y, however, are instantiated in the architecture. Through the port map aspects, the two adder-instantiation statements specify the connectivities of the two instances. The specifications of performance aspects, i.e., the delay time parameter values, are delayed until the overall design is completed and are defined through configurations Adderx and Addery.

The count of effective gates for the XY_Adders module totals 6548. With a utilization of 80%, the practical gate count is estimated to be 8185.

Table 5-7.   Functional Table of the Y Adder.

$y1$	$y2$	SUM_Y
0	0	0
0	1	If f = 1 then 0 else (Y_c) Bus_A
0	2	(Y_c) Bus_A
0	3	$(Y_c)\ 2^{-i}$ *X
1	0	SUM_Y
1	1	If f = 1 then SUM_Y else SUM_Y (Y_c) Bus_A
1	2	SUM_Y (Y_c) Bus_A
1	3	SUM_Y $(Y_c)\ 2^{-i}$ *X

* For the definition of Y_c see Table 5-8.

Table 5-8.   The Values of the X and Y Adders' Control Signals.

XY	X_c	Y_c
0	0	0
1	Ax	Ay
2	$flag$	$flag$
3	1	1

```
entity XY_Adders is
  generic (SgnY_TP_Delay : TIME := 1 ns);
  port (Data_A, Data_B : in BIT_32;
        X : out BIT_32;
        X_state : out VIRTUAL_SIGNAL;
        sign_Y : buffer BIT;
        sign_y2 : out BIT;
        sft_i : in BIT_5;
        x1, x2, y1, T, Txy : in BIT;
        y2, XY : in VECT_2;
        Ax, Ay, f, en, phase_1, phase_2 : in BIT);
end XY_Adders;
```
(Next page)

Code 5-19. VHDL description of the XY_Adders.

```
 ─── (continued) ───
architecture Structural of XY_Adders is
  component Adder
    port (addend_1, addend_2 : in BIT_32;
          sum : out BIT_32 := 0;
          funct : in BIT;
          phase_1, phase_2 : in BIT);
  end component;
  signal SUM, SUM_X, SUM_Y, X_sft_in, Y_sft_in, X_sft_out, Y_sft_out : BIT_32;
  signal A_in, B_in, x1_in, x2_in, y1_in, y2_in, y21 : BIT_32;
  signal GND : BIT_32 := 0;
  signal num_of_bits : BIT_5;
  signal one : BIT := '1';
  signal Xc, Yc, Xc_1, Yc_1, dummy, zero : BIT;

begin
  In_A: Inbuf_32 port map (Data_A, A_in, phase_1);
  In_B: Inbuf_32 port map (Data_B, B_in, phase_1);
  M2_x1: Mux32_2 port map (GND, SUM_X, x1_in, x1);
  M2_x2: Mux32_2 port map (Y_sft_out, B_in, x2_in, x2);
  Add_X: Adder port map (x1_in, x2_in, SUM_X, Xc, phase_1, phase_2);
  M2_y1: Mux32_2 port map (GND, SUM_Y, y1_in, y1);
  M2_y21: Mux32_2 port map (A_in, GND, y21, f);
  M4_y2: Mux32_4 port map (GND, y21, A_in, X_sft_out, y2_in, y2);
  Sgn_of_y2: process (y2_in)
              begin
                if y2_in < 0 then
                  dummy <= '1';
                else
                  dummy <= '0';
                end if;
              end process Sgn_of_y2;
  L2_sgny2: Latch port map (dummy, phase_2, sign_y2);
  Add_Y: Adder port map (y1_in, y2_in, SUM_Y, Yc, phase_1, phase_2);
  Sgn_of_Y: process (SUM_Y)
              begin
                if SUM_Y < 0 then
                  sign_Y <= transport '1' after SgnY_TP_Delay;
                else
                  sign_Y <= transport '0' after SgnY_TP_Delay;
                end if;
              end process Sgn_of_Y;
  M4_Xc: Mux_4 port map (zero, Ax, f, one, Xc_1, XY);
  L2_Xc: Latch port map (Xc_1, phase_2, Xc);
  M4_Yc: Mux_4 port map (zero, Ay, f, one, Yc_1, XY);
  L2_Yc: Latch port map (Yc_1, phase_2, Yc);
  L2_sft: Latch_5 port map (sft_i, phase_1, num_of_bits);
  M2_SX: Mux32_2 port map (SUM_X, Data_A, X_sft_in, T);
  M2_SY: Mux32_2 port map (SUM_Y, Data_B, Y_sft_in, T);
  X_sfter: R_Shifter port map (X_sft_in, X_sft_out, num_of_bits);
  Y_sfter: R_Shifter port map (Y_sft_in, Y_sft_out, num_of_bits);
  M2_Sum: Mux32_2 port map (SUM_X, SUM_Y, SUM, Txy);
  T_X: Tristate_32 port map (SUM, X, X_state, en, phase_1);
end Structural;
```

Code 5-19. (cont.) VHDL description of the XY_Adders.

5.4.3.4   The Set_Flag Module

The Set_Flag module is the hardware implementation of a flag whose source is selected by the signal *setf* during phase-two. The circuit diagram of this module is shown in Figure 5-21 and the corresponding VHDL description is listed in Code 5-20. The value of the flag may be the exclusive-or of the two most significant bits of the Z adder output (*setf* = 3), the sign bit of the Z adder output (*setf* = 2), or the sign bit of the y addend (*setf* = 1). The flag value is ready to propagate at phase-one of the next cycle. This module requires 28 gates.

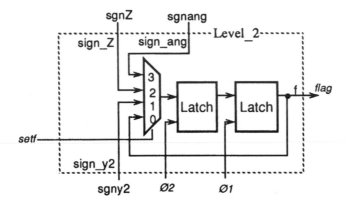

Figure 5-21.   Circuit diagram of the Set_Flag module.

```
entity Set_Flag is
  port (sign_Z, sign_y2, sign_ang : in BIT;
        phase_1, phase_2 : in BIT;
        setf : in VECT_2;
        f : out BIT);
end Set_Flag;

architecture Structural of Set_Flag is
  signal feedback, q1, q2 : BIT;
begin
  M4: Mux_4 port map (feedback, sign_y2, sign_Z, sign_ang, q1, setf);
  L2: Latch port map (q1, phase_2, q2);
  L1: Latch port map (q2, phase_1, feedback);
  f <= feedback;
end Structural;
```

Code 5-20. VHDL description of the Set_Flag module.

5.4.3.5 The ROM Module

The circuit diagram of the ROM module is shown in Figure 5-22 and the corresponding VHDL description is listed in Code 5-21. This module contains a

Level_3 entity of a 32-word, 32-bit, read-only memory unit (Mem_32). The memory unit stores 32 constants; the first 26 are used for the cordic calculation and the rest are for the IKS calculation. The address to the ROM comes from the control vector during normal operation and from the output of the counter within the cordic control unit during cordic operation. The output of this module is connected to the bus and originates from the Mem_32 unit as a tristate output. The effective and practical gate counts of this module are estimated to be 427 and 534, respectively.

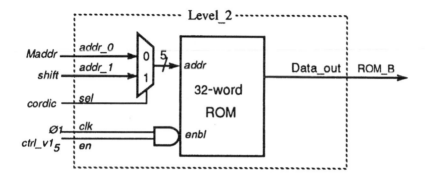

Figure 5-22.   Circuit diagram of the ROM.

```
entity ROM is
  generic (AND_Gate_Delay : TIME := 1.6 ns);
  port (addr_0, addr_1 : in BIT_5;
        sel, en, clk : in BIT;
        Data_out : out BIT_32;
        D_state : out VIRTUAL_SIGNAL);
end ROM;

architecture Structural of ROM is
  component Mem_32
    port (addr : in BIT_5;
          D_out : out BIT_32;
          D_state : out VIRTUAL_SIGNAL;
          en : in BIT);
  end component;
  signal enbl : BIT;
  signal addr : BIT_5;

begin
  enbl <= transport en and clk after AND_Gate_Delay;
  M2_addr: Mux5_2 port map (addr_0, addr_1, addr, sel);
  Cons: Mem_32 port map (addr, Data_out, D_state, enbl);
end Structural;
```

Code 5-21. VHDL description of the ROM.

5.4.3.6 The J Register Module

The J Register is the only module accessible by the user for I/O during the chip's normal operation. As shown in Figure 5-23, it is built around a 16-word register file, which is specified as a Level_3 entity Reg in the VHDL description (Code 5-22). The control signal *sel* is used to multiplex the two sets of addresses and read/write signals. During computation (*sel*=1), the control and address signals from the chip's control store are used. Otherwise, those from the chip's interface (*Waddr*, *Ed*, and *r_w*) are used. The *clr* signal is used in conjunction with the address signal to reset individual register words during testing. (For the assignment of the register's words, see Section 4.1.1.) The count of effective gates for the J Register module is 5857. The practical gate count is estimated to be 7321 given a utilization of 80%. This figure is derived based on a non-optimized implementation of the 16-word register sub-module. The actual gate count is likely to be smaller as there appears to be much room for minimization.

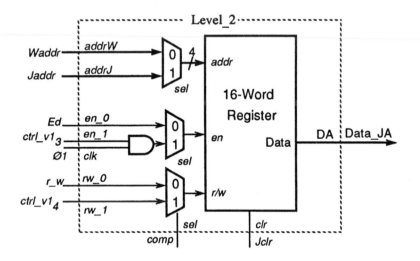

Figure 5-23. Circuit diagram of the J Register.

The 16-Word Register File

The circuit diagram of one possible implementation of a register file containing 16 words for use in the J Register module is shown in Figure 5-24. The *jth* bit cell of each word contains a latch whose output is connected to the *jth* bit of the data bus via a TG (Transmission Gate) controlled by the word enable signal. The data bus will be precharged during phase-two and thus the enable signal must remain low at phase-two (by the external circuitry). The *ith* word enable signal, when turned on after decoding during phase-one, will turn on all the TGs in the

```
entity J_Register is
  generic (AND_Gate_Delay : TIME := 1.6 ns);
  port (addrW, addrJ : in BIT_4;
        rw_0, rw_1, en_0, en_1, sel, clk, clr : in BIT;
        DA : buffer BIT_32;
        DAin : in BIT_32;
        DAsto : out VIRTUAL_SIGNAL);
end J_Register;

architecture Functional of J_Register is
  component Reg
    port (addr : in BIT_4;
          r_w, en, clr : in BIT;
          Data : buffer BIT_32;
          D_in : in BIT_32;
          state : out VIRTUAL_SIGNAL);
  end component;
  signal reg_rw, en, enJ : BIT;
  signal addr : Bit_4;

begin
  M2_addr: Mux4_2 port map (addrW, addrJ, addr, sel);
  M2_rw: Mux_2 port map (rw_0, rw_1, reg_rw, sel);
  M2_en: Mux_2 port map (en_0, enJ, en, sel);
  enJ <= transport en_1 and clk after AND_Gate_Delay;
  J: Reg port map (addr, reg_rw, on, clr, DA, DAin, DAsto);
end Functional;
```

Code 5-22. VHDL description of the J Register.

ith word. The *r/w* signal will determine whether the data is read from or written to the latch. The circuit developed here provides an upper-bound estimate of the area and read/write time parameters. The practical gate count of this sub-module is estimated to be 5757 for a utilization of 90%.

5.4.3.7 The R Register Module

The circuit diagram and the functional table of the R Register are shown in Figure 5-25. It is constructed using an 8-word, 2-port register file. The 8-word register file is similar to the 16-word file of the J Register module. The only modification is the addition of a read-only port with its own address. For data coherency, the write (via Port_A) and read (via Port_B) operations are not allowed to occur at the same word simultaneously. There are two possible sources during a write operation: Bus A and the output of the Z adder. The connection from the Z adder is a direct link through a tristate buffer which will be at high impedance when the data source of the write operation is from the bus. Transmission gates are used to connect the register file's data I/O port to the chip's bus so that the bus can be isolated from the I/O port when the data source of the write operation is the Z adder.

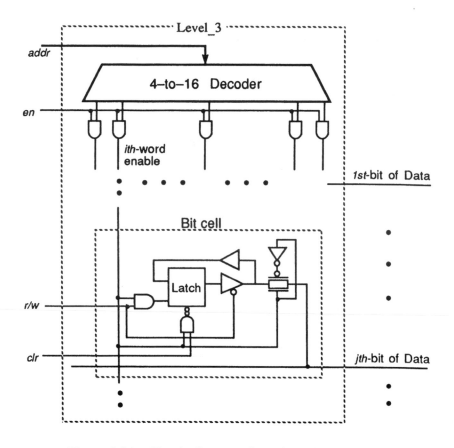

Figure 5-24. Circuit diagram of a 16-word register file.

The VHDL description of the R Register is listed in Code 5-23. Since the internal bus (denoted by an asterisk in the circuit diagram) has three potential drivers, the register output (PA_Data), the direct link from the Z adder (Z_in), and one end of the transmission gate (S0), the value of the internal bus needs to be resolved. This is done by two process statements, one for handling the state transitions and the other for the steady states. In addition, another process statement is included to pass the updated internal bus value to all reader signals at steady state. To emphasize the fact that each of these three process statements implements a facet of the bus resolution process, they are grouped together and bounded by the block statement Resolve.

The process statement that implements the state transitions (ST) is listed in Code 5-24. It performs two basic functions of constraint violation checks and execution of the connecting protocol to the TG. One obvious constraint is that no more than one driver should be present simultaneously. Another constraint forbids the condition of reading without a driver. It is conceivable, however,

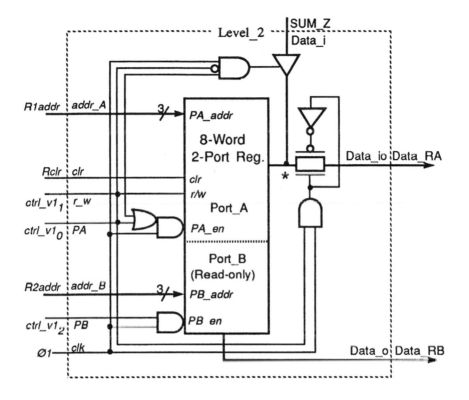

Operation: when $\varnothing 1=0$, all outputs are in HiZ-state;
when $\varnothing 1=1$, use the following table:

PA	r_w	Operation (Port_A)
0	0	disable
0	1	Reg(*addr_A*) ← Data_i
1	0	Data_io ← Reg(*addr_A*)
1	1	Reg(*addr_A*) ← Data_io

PB	Data_o
0	HiZ
1	Reg(*addr_B*)

Figure 5-25. Circuit diagram and functional table of the R Register.

that a reading signal may arrive earlier than the driving signal during the transition. Consequently, the ST statement incorporates a timer mechanism (for simulation only) so that the condition of reading without a driver is tolerated for a specified period.

The effective and estimated practical gate counts of the R Register module are 3936 and 4920, respectively.

```
entity R_Register is
  generic (BUFFER_Delay : TIME := 1 ns;
           AND_Gate_Delay : TIME := 1.6 ns;
           INV_AND3_Delay : TIME := 1.9 ns;
           Data_Valid : TIME := 0.2 ns;
           OR_Gate_Delay : TIME := 2.1 ns;
           Leakage_Time : TIME := 5 ns);
  port (addr_A, addr_B : in BIT_3;
        r_w, PA, PB, clk, clr : in BIT;
        D1_i : in BIT_32;
        D2_o : buffer BIT_32;
        D2_i : in BIT_32;
        D2sti : in VIRTUAL_SIGNAL;
        D2sto : out VIRTUAL_SIGNAL;
        D3_o : out BIT_32;
        D3sto : out VIRTUAL_SIGNAL);
end R_Register;
architecture Functional of R_Register is
        • • • (Appropriate signal declarations)
  component Reg
    port (PA_addr, PB_addr : in BIT_3;
          r_w, PA_en, PB_en, clr : in BIT;
          PA_Data : buffer BIT_32;
          PA_in : in BIT_32;
          PA_state : out VIRTUAL_SIGNAL;
          PB_Data : out BIT_32;
          PB_state : out VIRTUAL_SIGNAL);
  end component;
begin
  PA_rw <= transport r_w or PA after OR_Gate_Delay;
  PA_enable <= transport PA_rw and clk after AND_Gate_Delay;
  PB_enable <= transport PB and clk after AND_Gate_Delay;
  R: Reg port map (addr_A, addr_B, r_w, PA_enable, PB_enable, clr,
                   PA_Data, PA_in, PA_st, D3_o, D3sto);
  read_z <= transport (r_w and (not PA)) and clk after INV_AND3_Delay;
  src <= D1_i after Data_Valid;
  Tri_st: process (src, read_z)
        begin
          if read_z = '1' then
            Z_in <= transport src after BUFFER_Delay;
            Z_st <= 1;
          else
            Z_st <= 0;
          end if;
        end process Tri_st;
  TG_ctrl <= transport clk and PA after AND_Gate_Delay;
  TG_A: T_Gate port map (S0, D2_o, S0_in, D2_i, S0sto,
                D2sto, S0sti, D2sti, TG_ctrl);
  Resolve: block
        • • • (Code that resolves the internal bus value)
  end block Resolve;
end Functional;
```

Code 5-23. VHDL description of the R Register.

```
ST: process (P1_st, S0sto, Z_st, timeout)
       variable sp1, sa, sz : VIRTUAL_SIGNAL;
       variable status : VIRTUAL_SIGNAL;
       variable new_value : BIT_32;
   begin
   assert (timeout = '0') or timeout'STABLE
      report "Time out for reading without driver"
      severity WARNING;
   sp1 := P1_st;
   sa := S0sto;
   sz := Z_st;
   status := 0;      -- status counts the active drivers
   if sp1 = 1 then status := status + 1; end if;
   if sa = 1 then status := status + 1; end if;
   if sz = 1 then status := status + 1; end if;
   assert status <= 1
      report "more than one driver is active"
      severity ERROR;
   if status = 0 then    -- No driver
      if sa = -1 then status := -1; end if;
      if sp1 = -1 then status := -1; end if;
   end if;
   if sa = 2 then        -- TG_A requests status
      S0sti <= status;
   elsif (status /= 1) and (sa = -1) then   -- TG_A is reading
      S0sti <= 0;    -- The previous driver has dropped
   elsif (sa = 0) and (not S0sto'stable) then   -- TG_A open
      S0sti <= 0;
   end if;
   if status = -1 then
      if start_timer = '0' then start_timer <= '1'; end if;
   else
      if start_timer = '1' then start_timer <= '0'; end if;
   end if;
   end process ST;
```

Code 5-24. The Process statement resolving the internal bus value.

5.4.3.8 The Cordic Control Module

The circuit diagram of the Cordic Control unit is shown in Figure 5-26 and the corresponding VHDL description is listed in Code 5-25. For simplicity, the generic parameters and signal declarations are not listed. This module performs two functions. The first is to generate the adder control signals from the sign bits of the Z adder and Y adder according to the cordic algorithm as specified in Figure 4-11. The second function involves transferring the control from and to the Main Control unit. This is accomplished as follows. When m is non-zero, the signal *cordic* will be raised to the high state at phase-two, which will inhibit the advancement of the program counter (within the Main Control unit). Meanwhile, when either m is non-zero or *setf* is equal to 2, the phase-one clock

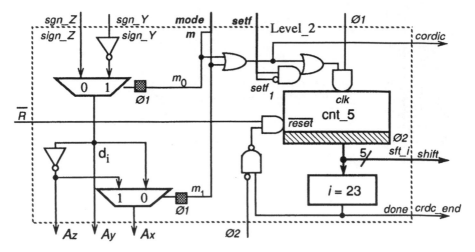

Figure 5-26. Circuit diagram of the Cordic Control.

```
entity Cordic_Ctrl is
  generic ( . . . );
  port (m, setf : in VECT_2;
        sign_Z, sign_Y, R_bar, phase_1, phase_2 : in BIT;
        Ax, Ay, Az : out BIT;
        cordic, done : buffer BIT;
        sft_i : buffer BIT_5);
end Cordic_Ctrl;
architecture Logic of Cordic_Ctrl is
          • • •  (Appropriate signal declarations)
begin
  sign_Y_neg <= transport not sign_Y after INV_Delay;
  L1_sel0: Latch port map (m(0), phase_1, sel0);
  L1_sel1: Latch port map (m(1), phase_1, sel1);
  M2_di: Mux_2 port map (sign_Z, sign_Y_neg, di, sel1);
  Az <= s1;   s1 <= transport not di after INV_Delay;
  Ay <= di;
  M2_Ax: Mux_2 port map (di, s1, Ax, sel0);
  cordic <= transport m(1) or m(0) after OR_Gate_Delay;
  sf2 <= transport setf(0) and (not setf(1)) after INV_AND_Delay;
  cnt_on <= transport sf2 or cordic after OR_Gate_Delay;
  s2 <= transport cnt_on and phase_1 after AND_Gate_Delay;
  Cnt_5: Counter generic map (5)
                  port map (reset_bar, s2, num);
  L2_num: Latch_5 port map (num, phase_2, sft);
  sft_i <= sft;
  reset <= transport done nand (not phase_2) after AND_Gate_Delay;
  reset_bar <= transport reset and R_bar after AND_Gate_Delay;
  done <= '1' after Compare_Time when sft_i = 23 else
          '0' after Compare_Time;
end Logic;
```

Code 5-25. VHDL description of the Cordic Control.

signal is allowed to trigger the counter within the cordic control unit. (The *setf* signal is used for starting the counter for mode 1 operation. This is because the cordic algorithm requires that this mode start counting from 1 instead of 0 as in the other two modes. The *setf* signal always occurs just before the mode 1 operation when its value is 2.) During the iteration, the counter output is used for both the shifter control and the ROM addresses for fetching the cordic constants. When the count reaches 23, the comparator will assert the signal *done*, which in turn resets the counter when the phase-two clock drops. The detailed timing diagram of the cordic control is shown in Figure 4-17.

The count of effective gates for this module is 131. The practical gate count is estimated to be 164, assuming a utilization of 80%.

5.4.3.9 The Main Control Module

The Main Control module controls the execution of the IKS microprogram and the interface to the user through the signals D, S, and R. The interface function implements the procedure specified in Figure 4-3. The module as a whole can be viewed as implementing the state machine of Figure 40. The chip is initially in the *Idle* state where input of the transformation matrix into the J Register may occur. A pulse applied to the signal S will put the chip in the *Run* state and start the computation. From this state, the chip may enter the *Stop* state when a cordic instruction (indicated by the non-zero value of *mode*) is encountered. The chip will return to the *Run* state when the cordic iteration ends. When the program counter (*pc*) reaches 127, the IKS computation is complete and the chip enters the *Done* state where output of the result may proceed. At any point in time, a pulse on the reset signal R will force the chip to reenter the *Idle* state.

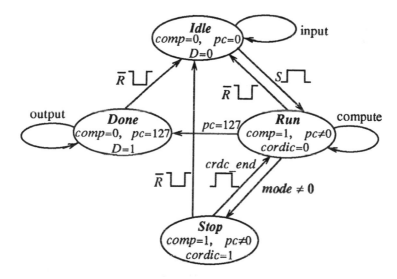

Figure 5-27. The state diagram of the IKS chip.

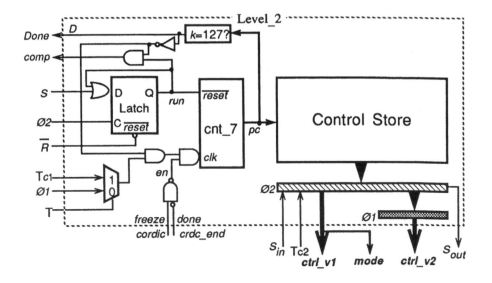

Figure 5-28. Circuit diagram of the Main Control.

The circuit diagram of the Main Control module is shown in Figure 5-28. The 7-bit counter in the module serves as a program counter pointing to the next MACC instruction in the Control Store. Before the computation starts, the input to the reset port of the counter (*run*) will remain low. The assertion of the signal *S* will be sensed at phase-two causing the signal *run* to rise. This enables the phase-one signal to increment the counter since the *en* signal is normally high. The instruction is latched at phase-two into a 40-bit wide register with scan-input capability after decoding. This control vector is split into two parts as was shown in Figure 4-21. The first part includes all control signals affecting phase-one events. The second part contains all control signals that affect the phase-two events. They must remain stable until the phase-one of the next cycle (see Section 4.3.2.2.2). When the fetched instruction is a cordic operation, the non-zero value of the signal *mode* will cause the Cordic Control unit to raise the *cordic* signal to high. The *en* signal will then become low and block the phase-one clock from advancing the counter. The *en* signal will remain high until the *crdc_end* signal from the Cordic Control unit becomes high, indicating that the cordic iteration has finished. The program counter will then resume counting from the next cycle. The counter also stops when its output reaches 127, i.e., the end of the IKS microprogram, and the signal from the comparator is connected to the chip's external signal *D* to indicate the end of the IKS computation. During testing, the clock input to the counter is directly under the user's control. Another test clock input (Tc2) is used to control the shifting operation of the control vector latch.

```
entity Main_Ctrl is
  generic (  .  .  . );
  port (done, freeze, S, R_bar, T : in BIT;
        phase_1, phase_2, T_c1, T_c2, s_in1, s_in2 : in BIT;
        D : buffer BIT;
        comp, s_out1, s_out2 : out BIT;
        mode : out VECT_2;
        Jaddr : out BIT_4;
        R1addr : out BIT_3;
        Maddr : out BIT_5;
        ctrl_v1 : out C_VECTOR_1;
        ctrl_v2 : out C_VECTOR_2);
end Main_Ctrl;

architecture Logic of Main_Ctrl is
            • • • (Appropriate signal declarations)
  component C_Store
    port (p_cnt : in BIT_7;
          ctrl_v1 : out PLA32_WORD;
          ctrl_v2 : out PLA8_WORD;
          addr : out BIT_24);
  end component;
begin
        -- Interface Logic
  start <= transport run or start after OR_Gate_Delay;
  L2_run: Latch_r port map (start, phase_2, R_bar, run);
  Cnt_7: Counter generic map (7)
                  port map (run, clk_in, pc);
  clk_in <= transport en and clk after AND_Gate_Delay_1;
  en <= transport freeze nand (not done) after INV_NAND_Delay;
  clk <= transport D_bar and clk_src after AND_Gate_Delay_2;
  M2_clk: Mux_2 port map (phase_1, T_c1, clk_src, T);
  D <= '1' after Compare_Time when pc = 127 else
        '0' after Compare_Time;
  D_bar <= transport D after INV_Delay;
  comp <= transport D_bar and run after AND_Gate_Delay_3;
        -- Control Signal Generation
  CS: C_Store port map (pc, pla_v1, pla_v2, addrs0);
  ph1_signal <= pla_v1 (0 to 10);
  ph2_signal <= pla_v1 (11 to 19) & pla_v2;
  L1_C1: Latch_s1 port map (ph1_signal, phase_2, ctrl_v1, T_c2, s_in1, s_out1);
  L1_adrs1: Latch_24 port map (addrs0, phase_2, addrs1);
  L1_v2: Latch_s2 port map (ph2_signal, phase_2, v2, T_c2, s_in2, s_out2);
  L2_C2: Latch_C2 port map (v2, phase_1, ctrl_v2);
  unpack: process (addrs1)
                  • • •
          end process unpack;
end Logic;
```

Code 5-26. VHDL description of the Main Control.

The main content of the VHDL description of the Main Control module is listed in Code 5-26. A Level_3 entity (C_Store) is instantiated in the architecture body. The architecture description can be partitioned into two parts of interface logic and the control signal generation. There are three address fields in the microcode. The three addresses are packed together as an integer stored in the ROM. The Control Store passes the packed addresses out essentially untouched. The process Unpack is used to translate the number back into the three addresses of the microcode.

The Main Control module requires 2999 effective gates, or a practical gate count of 3749 assuming 80% utilization.

The Control Store Sub-module

The circuit diagram of the Control Store is shown in Figure 5-29. It consists of a 128-word, 22-bit ROM and a decoder. Each word is a microcode whose format was defined in Figure 4-21. The decode sub-module takes the first eight bits of the microcode (opcode1 and opcode2) as input and decodes them into the control signals according the code maps specified in Appendix D.

The VHDL description of this Level_3 circuit is listed in Code 5-27. The architecture body contains three macrocells that are used only by this module. For this reason, their declarations have not been put in the package Cells. The microcodes stored in the entity Codes are represented as integers. An Unpack process translates the integer number into the three fields of opcode1, opcode2, and addresses.

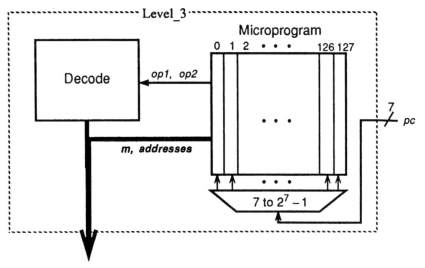

Figure 5-29. Circuit diagram of the Control Store sub-module.

```
entity Ctrl_Store is
  port (p_cnt : in BIT_7;
        ctrl_v1 : out PLA32_word;
        ctrl_v2 : out PLA8_word;
        addr : out BIT_24 := 0);
end Ctrl_Store;
architecture Dataflow of Ctrl_Store is
            • • • (Appropriate signal declarations)
  component Codes
    port (pc : in INTEGER;
          microinst : out INTEGER);
  end component;
  component Decoder_1
    port (opcode : in BIT_5;
          pla_out : out PLA32_WORD);
  end component;
  component Decoder_2
    port (opcode : in BIT_3;
          pla_out : out PLA8_WORD);
  end component;
begin
  M_Prog: Codes port map (p_cnt, m_inst);
  D_1: Decoder_1 port map (opcode1, ctrl_v1);
  D_2: Decoder_2 port map (opcode2, ctrl_v2);
  Unpack: process (m_inst)
              constant field1 : INTEGER := 2 ** 17;
              constant field2 : INTEGER := 2 ** 14;
              variable dummy : INTEGER;
            begin
              dummy := m_inst;
              opcode1 <= dummy / field1;
              dummy := dummy rem field1;
              opcode2 <= dummy / field2;
              addr <= dummy rem field2;
            end process Unpack;
end Dataflow;
```

Code 5-27. VHDL description of the Control Store.

     The design entity that stores the microcodes is named M_Code and has been
analyzed into the library Macros. This design entity has two alternative
architecture bodies: one called IKS and the other called Instruction_Set. After
the Ctrl_Store has been analyzed into the Level_3 library, two configurations
are created by binding the Codes instance M_Prog to these two architectures of
M_Code. The VHDL program that creates the IKS_Store as a configuration of
the Ctrl_Store is listed in Code 5-28. Once the two configurations are analyzed
into the Level_3 library, two configurations of the Main Control entity can be
created by binding the C_Store instance CS to either of the IKS_Store or the
Instruction_Set_Store. Code 5-29 shows the VHDL program that configures the

```
configuration IKS_Store of WORK.Ctrl_Store is
  for Dataflow
    for M_Prog : Codes
      use entity Macros.M_Code (IKS)
        generic map (  .  .  . );
    end for;

    for D_1 : Decoder_1
      use entity Macros.Codemap_1 (PLA_32)
        generic map (  .  .  . );
    end for;

    for D_2 : Decoder_2
      use entity Macros.Codemap_2 (PLA_8)
        generic map (  .  .  . );
    end for;
  end for;
end IKS_Store;
```

Code 5-28. The VHDL program to create the IKS_Store
as a configuration of the Ctrl_Store.

```
library IKS_Chip, Macros, Level_3;
use IKS_Chip.Cells.All, Macros.ALL, Level_3.ALL;

configuration M_Ctrl_IKS of WORK.Main_Ctrl is
  for Logic
    for CS: C_Store
      use configuration Level_3.IKS_Store;
    end for;

    for ALL : Counter
      use entity Macros.Counter (Behavioral);
        generic map (  .  .  . );
    end for;

      • • •  (other configuration items)

  end for;
end M_Ctrl_IKS;
```

Code 5-29. The VHDL program to configure the Main_Ctrl using
the library unit IKS_Store for its C_Store instance.

C_Store containing the codes for the IKS computation. Note that in the binding indication for the CS instance, instead of the usual "entity (architectue)" pair, the configuration of IKS_Store is selected.

The practical gate count for the microprogram storage is about 1408. If the decoding is implemented with two PLAs, about 226 actual gates will be required. As a result, the practical gate count of the Control Store is estimated to be 2454.

5.4.4 The Level_1 Library

The Level_1 library contains two design entities: Datapath and Control. The Control entity is simply an assembly of the two Level_2 design units in the control category — the Main and Cordic Control units. The VHDL description of the Control entity is listed in Code 5-30. Note that prior to the analysis of this entity, the package Modules, which contains all templates (component declarations that have the same port interface lists as the entity declarations) of Level_2 units, has been analyzed into the Level_2 library.

After the analysis of the entity Control into the Level_1 library, two library units can be configured from the Control unit. The one that binds the instance MC to the configuration M_Ctrl_IKS, which has the IKS program in its Control Store and resides in Level_2 library, is named Ctrl_IKS. The other that binds MC to the configuration M_Ctrl_Instruction_Set, is named Ctrl_Instruction_Set.

```vhdl
library IKS_Chip, Level_2;
use IKS_Chip.Custom_Types.ALL, Level_2.Modules.ALL;

entity Control is
  port (S, R_bar, Test, T_clk1, T_clk2 : in BIT;
        Scan_1, Scan_2, phase1, phase2, sgn_Z, sgn_Y : in BIT;
        S_out1, S_out2, comp, Ax, Ay, Az : out BIT;
        Done, cordic : buffer BIT;
        shift : buffer BIT_5;
        ctrl_v1 : out C_VECTOR_1;
        ctrl_v2 : out C_VECTOR_2 := "000110000000000";
        Jaddr : out BIT_4;
        R1addr : out BIT_3;
        Maddr : out BIT_5);
end Control;

architecture Structural of Control is
  signal v2 : C_VECTOR_2;
  signal mode, setf : VECT_2;
  signal crdc_end : BIT;
begin
  MC: Main_Ctrl port map
                  (crdc_end, cordic, S, R_bar, Test,
                   phase1, phase2, T_clk1, T_clk2, Scan_1,
                   Scan_2, Done, comp, S_out1, S_out2,
                   mode, Jaddr, R1addr, Maddr, ctrl_v1, v2);
  CC: Crdc_Ctrl port map
                  (mode, setf, sgn_Z, sgn_Y, R_bar, phase1,
                   phase2, Ax, Ay, Az, cordic, crdc_end, shift);
  setf <= v2(13 to 14);
  ctrl_v2 <= v2;
end Structural;
```

Code 5-30. VHDL description of the Control.

The rest of the Level_2 entities, i.e., entities in the categories of functional unit and storage, are likewise combined to form the Datapath. The VHDL description of the Datapath is listed in Code 5-31.

```
entity Datapath is
  generic (Allowed_Time : TIME := 10 ns);   -- for reading w/o drivers
  port (ctrl_v1 : in C_VECTOR_1;
        ctrl_v2 : in C_VECTOR_2;
        phase1, phase2, Ed, r_w, cordic, comp : in BIT;
        Ax, Ay, Az, Jclr, Rclr, T, Txy : in BIT;
        Waddr, Jaddr : in BIT_4;
        R1addr : in BIT_3;
        Maddr, shift : in BIT_5;
        Bus_A, Bus_B : buffer BIT_32 := 0;
        Ain, Bin : in BIT_32;
        AWsti, BWsti : in VIRTUAL_SIGNAL;
        AWsto, BWsto : out VIRTUAL_SIGNAL := 0;
        sgnY : buffer BIT;
        sgnZ : out BIT);
end Datapath;

architecture Structural of Datapath is
        • • • (Appropriate signal declarations)
begin
  ROM_32: ROM port map (Maddr, shift, cordic, ctrl_v1(5),
                        phase1, ROM_B, ROMst);
  J: Reg_J port map (Waddr, Jaddr, r_w, ctrl_v1(4), Ed, ctrl_v1(3),
                     comp, phase1, Jclr, Data_JA, Bus_A, JAsto);
  R: Reg_R port map (R1addr, R2addr, ctrl_v1(1), ctrl_v1(0),
                     ctrl_v1(2), phase1, Rclr, SUM_Z, Data_RA,
                     Bus_A, RAsti, RAsto, Data_RB, RBsto);
  Mply : Mpyer port map (Bus_A_24, Bus_B_24, PROD, Data_MA, MAsto,
                     ctrl_v1(8), phase1, phase2);
  Add_Z: Zadder port map
                     (Bus_A, Bus_B, PROD, SUM_Z, Data_ZA, ZAsto, signZ,
                     sgnang, oprndz1, oprndz2, op_Zadd, op_zang,
                     flag, Az, ctrl_v1(7), phase1, phase2);
  Add_XY: XYadders port map
                     (Bus_A, Bus_B, Data_XA, XAsto, sgnY, sgny2, shift,
                     ctrl_v2(0), ctrl_v2(1), ctrl_v2(2), T, Txy, oprndy2,
                     op_XY, Ax, Ay, flag, ctrl_v1(6), phase1, phase2);
  FB: SetFlag port map (signZ, sgny2, sgnang, phase1, phase2, setf, flag);
  sgnZ <= signZ;
  R2addr <= Maddr rem 8;
  op_zang <= ctrl_v1(9 to 10);
  oprndy2 <= ctrl_v2(3 to 4);
  oprndz1 <= ctrl_v2(5 to 6);
  oprndz2 <= ctrl_v2(7 to 8);
  op_Zadd <= ctrl_v2(9 to 10);
  op_XY <= ctrl_v2(11 to 12);
  setf <= ctrl_v2(13 to 14);
```
───────── (Next page) ─────────

Code 5-31. VHDL description of the Datapath.

```
┌──────── (continued) ─────────┐
│ Trunc_24_A: process (Bus_A)
│             begin
│                 Bus_A_24 <= Bus_A / 256;
│             end process Trunc_24_A;
│ Trunc_24_B: process (Bus_B)
│             begin
│                 Bus_B_24 <= Bus_B / 256;
│             end process Trunc_24_B;
│
│ Resolve_B: block
│             signal timeout, start_timer : BIT;
│             begin
│                     -- State Transition for Bus B
│             ST_B: process (BWsti, ROMst, RBsto, timeout)
│                         . . .
│                     end process ST_B;
│                     -- Steady States for Bus B
│             SS_B: process (Data_RB, RBsto, ROM_B, ROMst, Bin, BWsti)
│                         . . .
│                     end process SS_B;
│             Timer: process (start_timer)
│                         . . .
│                     end process Timer;
│             end block Resolve_B;
│
│ Resolve_A: block
│             signal timeout, start_timer : BIT;
│             begin
│                     -- State Transition for Bus A
│             ST_A: process (AWsti, MAsto, ZAsto, XAsto, JAsto, RAsto, timeout
│                         Data_XA, XAsto, Data_JA, JAsto, Data_RA, RAsto)
│                         . . .
│                     end process ST_A;
│                     -- Steady States for Bus A
│             SS_A: process (Ain, AWsti, Data_MA, MAsto, Data_ZA, ZAsto,
│                         Data_XA, XAsto, Data_JA, JAsto, Data_RA, RAsto)
│                         . . .
│                     end process SS_A;
│             Timer: process (start_timer)
│                         . . .
│                     end process Timer;
│             end block Resolve_A;
│ end Structural;
└──────────────────────────────┘
```

Code 5-31. (cont.) VHDL description of the Datapath.

The architecture body of the Datapath is composed of two parts. The first part (page 164) deals with connecting various Level_2 instances. The second part (page 165) contains process statements that simulate hardware truncation (the two Trunc_24 statements) and resolve bus signal values. Similar to the bus resolution scheme described in Section 5.4.3.7, the value of each bus in the Datapath is resolved by a **block** of process statements that handle the state transitions and steady states separately. The state transition process (ST) performs constraints violation checks and executes the connecting protocol to the TGs. The steady states processes (SS) are responsible for passing the values from the active driver to the bus. In addition, each block also has a process statement implementing a timer mechanism to allow the condition of reading without a driver to be tolerated for a specified time period during the transition.

There are three potential drivers to Bus B. They are one of the chip's data I/O (Bin), the ROM output (ROM_B), and the Port_B of the R Register (Data_RB). The connection to the chip's data I/O is bidirectional and is made through TGs. Since the TG's connection protocol has been defined (see Section 5.4.2.7), the provision of I/O ports that conform to the data structure and signal naming conventions of the TG is sufficient for the bus resolution scheme to incorporate code executing the TG's connecting protocol. This allows the instantiation of the TGs to be delayed until the next level so that the complexities of the description, and thus the complexities of the subsequent testing and debugging, are reduced. The state transition statement for Bus B (ST_B) monitors the changes of the virtual_signals associated with these three potential driver signals as indicated in the sensitivity list of the statement ST_B. The steady state process statement SS_B responds to the active driver and passes its value to the signal Bus_B.

The block statement for Bus A has a similar structure to that of Resolve_B. The only difference is that there are three more potential driver signals and two of them are bidirectional (the connections to the J and R Registers). This difference is reflected in the longer sensitivity lists of the two process statements.

5.4.5 The Root Library

The root of the IKS chip circuit hierarchy is described by the entity IKSChip, which is analyzed into the root library IKS_Chip. It essentially connects the two Level_1 entities of Datapath and Control together and instantiates two TG instances for interfacing the chip's data I/O to the chip's two busses. Note that the port interface list of the entity declaration is the same as presented in Figure 4-1 except that two auxiliary inputs (Bin and Din) and four virtual_signals are present due to the bidirectional nature of the two data I/O ports. The VHDL description of the IKSChip is listed in Code 5-32.

```
library IKS_Chip, Level_1;
use IKS_Chip.Custom_Types.ALL, IKS_Chip.Cells.T_Gate, Level_1.Modules.ALL;

entity IKSChip is
  generic (OR_Gate_Delay : TIME := 1 ns;
           TP_Delay : TIME := 0.5 ns);
  port (Addr : in BIT_4;
        arm_conf : in BIT_3;    -- not implemented
        r_w, R, S, Ed, Jclr, Rclr : in BIT;
        clk1, clk2, T, Tc1, Tc2, Txy, Sin, Ea, Eb : in BIT;
        Done : buffer BIT;
        Sout : out BIT;
        D, B : buffer BIT_32;
        Din, Bin : in BIT_32;
        Dstatoo, Bstateo : out VIRTUAL_SIGNAL := 0;
        Dstatei, Bstatei : in VIRTUAL_SIGNAL);
end IKSChip;

architecture Structural of IKSChip is
  signal ctrlv1 : C_VECTOR_1;
  signal ctrlv2 : C_VECTOR_2;
  signal cordic, comp, Ax, Ay, Az, sign_Y : BIT;
  signal sign_Z, Sout1, ctrlA, ctrlB : BIT;
  signal Jaddr : BIT_4;
  signal R1addr : BIT_3;
  signal Maddr, shift : BIT_5;
  signal BusA, BusB, BusAin, BusBin : BIT_32;
  signal BusAsti, BusAsto, BusBsti, BusBsto : VIRTUAL_SIGNAL;
begin
  DP: Dpath port map
              (ctrlv1, ctrlv2, clk1, clk2, Ed, r_w, cordic, comp,
               Ax, Ay, Az, Jclr, Rclr, T, Txy,  Addr, Jaddr, R1addr,
               Maddr, shift, BusA, BusB, BusAin, BusBin,
               BusAsti, BusBsti, BusAsto, BusBsto, sign_Y, sign_Z);
  CC: Ctrl port map
          (S, R, T, Tc1, Tc2, Sin, Sout1, clk1, clk2, sign_Z,
           sign_Y, Sout1, Sout, comp, Ax, Ay, Az, Done, cordic,
           shift, ctrlv1, ctrlv2, Jaddr, R1addr, Maddr);
  TG_A: T_Gate port map
              (BusAin, D,
               BusA, Din,
               BusAsti, Dstateo,
               BusAsto, Dstatei, ctrlA);
  TG_B: T_Gate port map
              (BusBin, B,
               BusB, Bin,
               BusBsti, Bstateo,
               BusBsto, Bstatei, ctrlB);
  ctrlA <= transport Ea or Ed after OR_Gate_Delay;
  ctrlB <= transport Eb after TP_Delay;

end Structural;
```

Code 5-32. VHDL description of the IKSChip.

5.5 Simulation Results

This section describes the VHDL simulations of the IKS chip. The simulations are carried out in a bottom-up fashion. Building block circuits including controlled I/O buffers, latches, multiplexers, shifters, counters, registers, and adders are first tested. The testing typically involves random generation of a set of data as input to the design units. The simulated responses of the output signals are then compared to what is expected from the functional tables to verify the correctness of the program semantics.

With all building block circuits thoroughly tested, the Level_2 units are tested next. Level_2 units invariably have instantiated design entities (instances) of a lower level. All instances within a unit are unbound when the unit is analyzed into the library because of the stated programming strategy. The binding of these instances must be specified before the simulation can be conducted. This is accomplished by configurations, which use the default values for the delay time parameters of the lower level entities when the related information is incomplete. Similar to the building block circuits, these modules are tested for their functionality with random input data.

Following the testing of the Level_2 units, the behavior of the two Level_1 entities of Datapath and Control are simulated. The cordic operation and non-cordic operations of the Datapath are tested separately. With each MACC instruction verified, the entire chip is then simulated from input to output on realistic data sets. The details of the simulations of the cordic operation, the MACC instructions, and the IKS computation are presented in the following subsections.

5.5.1 Simulation of the Cordic Operation

The input and expected output of the three modes of cordic operations are tabulated in Table 5-9. As shown in the table, there are two tests on the mode 1

Table 5-9. Summary of the Cordic Operation Tests.

Mode	Input	Expected output
1	z1, z2, with z1 > z2 z1, z2, with z1 < z2	$X = sqrt(z1*z1 - z2*z2)$ $X = 0$
2	z (in 1st quadrant) z (in 2nd quadrant) z (in 3rd quadrant) z (in 4th quadrant)	$X = cosz > 0, Y = sinz > 0$ $X = cosz < 0, Y = sinz > 0$ $X = cosz < 0, Y = sinz < 0$ $X = cosz > 0, Y = sinz < 0$
3	x, y with x > 0, y > 0 x, y with x < 0, y > 0 x, y with x < 0, y < 0 x, y with x > 0, y < 0	$Z = arctan (y,x)$, Z in 1st quadrant $Z = arctan (y,x)$, Z in 2nd quadrant $Z = arctan (y,x)$, Z in 3rd quadrant $Z = arctan (y,x)$, Z in 4th quadrant

operation and four testings on each of the mode 2 and mode 3 operations. A C program is developed to generate the test data as follows. The program accepts three 32-bit integers as inputs to the X, Y, and Z adders and a number indicating the mode. It then executes the cordic algorithm as was specified in Figure 4-11 and provides the intermediate results (the output of the three adders) during each iteration. A VHDL test program is developed, which instantiates the test configurations of the Datapath and the Cordic_Ctrl as the UUT (Unit_Under_Test) and the same input data is applied. By comparing the VHDL simulation results with the C program output, the correctness of the design with respect to the cordic operation is verified.

In all cordic simulation cases, a clock with 100-ns cycle time is used. The phase-one clock turns on at the beginning of each cycle for 30 ns. The phase-two clock has the same width with a phase-lag of 50 ns. The values of the three adder outputs at the end of the on-state of the phase-one clock, together with their function control signals from the cordic control unit, are reported. Also reported are the values of signals *shift* (from the cordic control unit) and sgnang (from the Z_Angle module).

Figure 5-30 shows the VHDL simulation results* of the mode 2, second type, cordic operation. For this type of operation, the input (z) is an angle in the second quadrant. (In this particular case, the input number 1610612736 corresponds to an angle value of 135°.) The angle is transferred to the Z adder during the second cycle (100 ns). The third cycle is the preprocessing cycle where the main control unit is still active. During this cycle, the flag is set to the exclusive-or of the two most significant bits of the Z adder output. Since the Z adder input is taken from the output of the Z_Angle module with *zang* = 3, the input angle z is transformed to z' such that -90° ≤ z' < 90°. The cordic iteration starts at the clock cycle of 300 ns and continues until the signal *shift* reaches 23. During the iteration cycles, the signal *cordic* remains high, which inactivates the main control unit. The main control unit is reactivated after the signal *shift* changes to 23. The clock cycle of 2700 ns is the post-processing cycle. The values of the X and Y adders reported at 2729 ns, however, are the results of the previous cycle, i.e., cos(z') in X adder and sin(z') in Y adder. The post-processing in this case is to complement both the sinc and cosine values. The results are reported at 2829 ns, with sinz now in the X adder and cosz in the Y adder. For this example, the absolute values of sinz and cosz are the same and equal to 759250125. Hence, the computation achieves an accuracy of 23 bits.

5.5.2 Simulation of MACC Instruction Set

The cordic operation (not counting the pre- and post-processing), takes 24 actual clock cycles to complete, yet represents only one MACC instruction from the viewpoint of the main control unit. Among the 125 microcodes that constitute the entire IKS microprogram, the cordic operations take up only 14.

---

* The original programs are written in VHDL 7.2 and run on the Intermetrics' VHDL Build 3 Simulator installed on a MicroVax.

10-SEP-88 19:24:07   VHDL Simulation Report Generator [V 1.1]   Page 2

Time			Signal Names					
(ns)	SHIFT	SUM_X	AX	SUM_Y	AY	SUM_Z	AZ	SGNANG
29	0	0	'0'	0	'0'	0	'1'	'0'
129	0	0	'0'	0	'0'	0	'1'	'0'
229	0	0	'0'	0	'0'	1610612736	'1'	'1'
329	0	652032874	'0'	0	'1'	-536870912	'0'	'0'
429	1	652032874	'1'	-652032874	'0'	0	'1'	'0'
529	2	978049311	'0'	-326016437	'1'	-316933405	'0'	'0'
629	3	896545202	'0'	-570528764	'1'	-149474498	'0'	'0'
729	4	825229107	'0'	-682596914	'1'	-64469742	'0'	'0'
829	5	782566800	'0'	-734173733	'1'	-21802411	'0'	'0'
929	6	759623871	'0'	-758628945	'1'	-447946	'0'	'0'
1029	7	747770294	'1'	-770498067	'0'	10231892	'1'	'0'
1129	8	753789810	'1'	-764656112	'0'	4891647	'1'	'0'
1229	9	756776747	'1'	-761711621	'0'	2221484	'1'	'0'
1329	10	758264465	'1'	-760233542	'0'	886398	'1'	'0'
1429	11	759006880	'1'	-759493050	'0'	218854	'1'	'0'
1529	12	759377726	'0'	-759122442	'1'	-114918	'0'	'0'
1629	13	759192394	'1'	-759307836	'0'	51968	'1'	'0'
1729	14	759285082	'0'	-759215162	'1'	-31475	'0'	'0'
1829	15	759238744	'1'	-759261505	'0'	10246	'1'	'0'
1929	16	759261914	'0'	-759238335	'1'	-10614	'0'	'0'
2029	17	759250329	'0'	-759249920	'1'	-184	'0'	'0'
2129	18	759244537	'1'	-759255712	'0'	5031	'1'	'0'
2229	19	759247433	'1'	-759252816	'0'	2424	'1'	'0'
2329	20	759248881	'1'	-759251368	'0'	1121	'1'	'0'
2429	21	759249605	'1'	-759250644	'0'	470	'1'	'0'
2529	22	759249967	'1'	-759250282	'0'	145	'1'	'0'
2629	23	759250148	'0'	-759250101	'1'	-17	'0'	'0'
2729	0	759250058	'0'	-759250191	'0'	64	'1'	'0'
2829	0	759250191	'0'	-759250058	'0'	0	'1'	'0'

Figure 5-30. The simulation result of mode 2 cordic operation.

To control the testing complexity and facilitate the debugging process, it is desirable to also test other operations before the entire IKS computation is simulated. At this juncture, the power of casting the chip's behavior into the MACC instruction set becomes obvious. Not only does this substantially reduce the number of test cases, but more important, the high-level characterization of an instruction concisely specifies *what* is being tested and thus enables the testing effort to be conducted efficiently. In addition, the simulation results of the MACC instruction set can be used for production testing.

The UUT (in VHDL) for this simulation is composed of the Datapath and the Main Control unit. In the test bench program, the Datapath instance is bound to the same configuration used in the cordic operation simulation. The Main_Ctrl instance, on the other hand, is bound to the configuration M_Ctrl_Instruction_Set, which has the MACC instruction set instead of the IKS program stored in its program ROM.

The test bench program controlling the simulation process first initializes both the J register file and R register file contents to random numbers and then starts the instruction execution. Signals of functional module outputs, the two register I/O ports, the two busses, and other key control signals are reported at the end of the on-state of each clock phase. Using the MACC instruction set table as a guide, the values of relevant signals of each clock cycle are compared to expected values. This verification process serves to eliminate mistakes made in describing the connections or in the process of entering data such as in the code map.

5.5.3 Simulation of the IKS Computation

After the cordic operation and the execution of the MACC instruction set are verified, the entity IKSChip is ready for the final simulation. The UUT in the test bench program is a configuration called Prototype of the entity IKSChip. The Prototype configuration uses the same Datapath configuration as the one used in the previous two simulations. The Control instance (CC), however, is bound to the configuration Ctrl_IKS.

Figure 5-31 illustrates the flow of the simulation for verifying the architecture design of the IKS chip. A C program (DKS) is developed to compute the Cartesian position and orientation (T) of the PUMA robot manipulator for a given joint angle set (Θ) using floating-point numbers. The elements of the homogeneous transformation matrix, T, however, are cast into 32-bit integers conforming to the formats specified in Figure 4-9. Another C program (IKS) is developed to implement the pseudocode of Appendix B. This program accepts the input T and computes the IKS using fixed-point calculations. It also provides all the intermediate results of each step, which are necessary for the debugging of the design and the VHDL program. The correctness of the IKS C program is easily verified by comparing Θ and $\Theta1$. To reduce the possibility of undetected errors due to some data peculiarities, 10

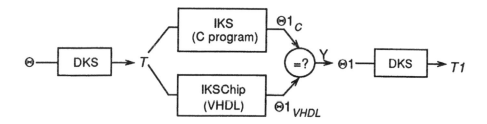

Figure 5-31. The flow of the IKS chip simulation.

joint angle sets are randomly generated and tested. The homogeneous transformation matrices are then used as input to the VHDL description of the IKS chip. The correctness of the chip design is verified by the exact match of $\Theta 1$s from the C program and from the VHDL simulation. Table 5-10 shows the errors of the IKS computation with respect to the original joint angle input in one such simulation.

Table 5-10. The Errors of the IKS Computation with Respect to the Joint Angles.

i	Θ (rad)	$\Theta 1$ (rad)	ε (rad)
0	0.180710	0.180710	0.000000
1	-0.074570	-0.074570	0.000000
2	0.488663	0.488664	0.000001
3	0.406296	0.406300	0.000004
4	0.006317	0.006317	0.000000
5	0.262054	0.262050	0.000004

Since in realistic situations the angle values are unknown, the errors of the computation should be assessed with respect to the position and orientation given. Therefore, the DKS program is again invoked to compute the position (the p vector) and orientation (the a and o vectors) from $\Theta 1$. This error in Cartesian coordinates for the same data set used in deriving the joint angle error of Table 5-10 is shown in Table 5-11. In this table, the last three elements are the

Table 5-11. The Error of the IKS Computation with Respect to the Position and Orientation.

	T	$T1$	ε
ox	0.138238	0.138238	0.000000
oy	-0.865541	-0.865541	0.000000
oz	-0.481383	-0.481383	0.000000
ax	0.208535	0.208536	0.000001
ay	0.500591	0.500591	0.000000
az	-0.840191	-0.840192	0.000001
px	135.461063	135.460621	0.000442
py	588.573889	588.572623	0.001266
pz	-185.233801	-185.233069	0.000732

position elements and have the units of mm. The results indicate that the errors in the orientation elements are negligible, and the errors in the position elements are well below 0.1 mm.

5.6 Summary

The two concepts of signal and entity are of fundamental importance in understanding how circuits are modeled in VHDL. Behavioral aspects — functions and timing — are high-level characterizations of the electrical states of wires and are specified through signal assignment statements. Structural aspects — the hierarchical relationships and connectivity among the building blocks — are more conveniently captured through design entities and their instantiations.

The potential of applying these concepts in ASIC processor designs has been demonstrated by the modeling of the IKS chip. VHDL models of basic building block circuits available from commercial gate array libraries have been presented. The IKS chip's architecture has been described following both the top-down approach of design decomposition and the bottom-up approach of simulation. The process of simulating the cordic operation, the MACC instruction set, and the IKS computation of the chip has been explained.

As an investigating vehicle, the IKS chip simulation project has produced fruitful results for modeling IC designs in VHDL. Specifically, techniques useful in modeling the transmission gates and bus-based communications have been described in terms of data structure and semantics. The library structure has been used to organize more than fifty library units in an intuitive way, which facilitates the recognition of the hierarchical relationships among these units. The use of data typing as an abstraction mechanism has been examined. A general timing model based on the timing description primitives provided by VHDL has been presented. A programming strategy based on the language construct configuration has been developed, which allows the delay time information to be specified in a progressive fashion. It not only saves programming and computation efforts, but also provides a structure for investigating the effects of using components of different performance characteristics and for simulation with back annotation after the physical design is complete. These results have laid a solid groundwork for VHDL-based behavioral modeling of IC architectures in the ASIC design environment.

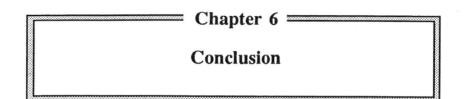

Chapter 6

Conclusion

Where a new system concept or new technology is used, one has to build a system to throw away, ...

Hence, plan to throw one away: you will, anyhow.

Frederick P. Brooks, Jr.
THE MYTHICAL MAN-MONTH: ESSAYS
ON SOFTWARE ENGINEERING (1975)

The major objective of this research is to obtain a better understanding of the interactions between the three domains of algorithm, architecture, and technology in the ASIC design environment through the development of a paradigm based on the architecture design of the IKS chip. This objective has been accomplished. Specifically, IC design knowledge has been organized into a conceptual framework. The process of deriving an architecture from an application algorithm has been formalized into an ASIC architecture design methodology. An architecture design paradigm based on an ASIC chip implementing the closed form IKS algorithm has been developed. Furthermore, the chip design has been documented in VHDL and its functionality has been verified by extensive VHDL simulations. The next section summarizes the salient results in these four areas. This is followed by a discussion of the implications of this work and future research issues.

6.1 Summary

To provide system designers with a logical view of the implementing technology, a conceptual framework for ASIC design has been described from a decision making perspective. The framework organizes a broad range of IC design knowledge into three frames of design process, design hyperspace, and design repertoire. Key concepts presented in the process frame include the hierarchical approach, the role of methodology, and a model for implementing design methodology in an integrated environment. The hyperspace frame articulates the design space concept and outlines the framing of the algorithm space and the architecture space. The repertoire collects various techniques for evaluating and optimizing the design alternatives. Under this framework, IC design activities are viewed as a process of making design decisions, and the emphasis is placed on the recognition and evaluation of design alternatives. The global strategy is to limit the search space through methodology implemented by CAE and KBES. Alternatives are recognized through a proper framing of the design space and optimized by various techniques.

Guided by this framework and based on the experience of the IKS chip design, the process of making architecture design decisions has been generalized into an ASIC architecture design methodology. The methodology partitions the architecture design process into three phases of functional unit configuration, communication configuration, and the control configuration, with each phase having its own decision focus. In the first phase of functional unit configuration, the focus is on finding an appropriate architecture style that best matches the characteristics of the task algorithm. Once the decision is made and fixed in the functional unit profile, the task algorithm is then translated into a pseudocode program. In the second phase of communication configuration, the data dependency inherent in the pseudocode program is unfolded in a dataflow table. The development of the system event timing model and the detailed designs of circuit modules proceed in parallel. From the desired dataflow pattern, an interconnection scheme evolves and is refined through manipulating design information of the dataflow table to negotiate a balance between maximization of dataflow efficiency and resources utilization. In the final phase of control configuration, the control signal patterns are analyzed to determine an encoding scheme. The control signals are also recast into an instruction set to facilitate the verification and testing process. Thus, embodied in this methodology is the new concept of derived instruction set computer (DISC) design, a manifestation of the basic principle of allocating hardware resources based on the tradeoff analysis of the application needs and the potential benefits in the algorithm-specific processor design environment.

The execution of the ASIC architecture design methodology is then illustrated by the design of a gate array chip implementing the closed form IKS algorithm for a robotic manipulator. The analysis of architectural alternatives has led to the idea of synthesizing an MAC structure with a cordic core as an on-chip coprocessor. The resultant architecture, called MACC for Multiplier Accumulator with a Cordic Core, can compute the IKS in 125 instruction cycles, or 445 clock cycles. The chip requires less than 100 I/O pins (including those for testing) and 50k gates even for a gate utilization as low as 60%. Based on the current gate array technology, the architecture can be implemented in a single chip. If the physical design of the chip can achieve a clock rate of 10 MHz, the IKS can be computed in less than 45 μs. Compared with a single cordic processor implementation, the area of MACC is estimated to be about

23% larger, but the computation time is reduced by more that one third, giving a overall area-time efficiency of 1.866 over the single cordic processor approach.

The complete design of the IKS chip architecture has been documented in VHDL and the correctness of the design has been verified by extensive simulations. The project has produced fruitful results for modeling IC designs in VHDL. Specifically, a set of techniques useful in modeling transmission gates and bus-based communications has been described. The library structure has been used to organize more than fifty library units in an intuitive way, which greatly facilitates the recognition of the hierarchical relationships among these units. The use of data typing as an abstraction mechanism has been examined. A general timing model based on the timing description primitives provided by VHDL has been presented. A programming strategy based on the language construct configuration has been developed, which strives to manage the delay time information in a systematic manner. It not only saves programming and computation efforts, but also provides a structure for investigating the effects of using components of different performance characteristics and for simulation with back annotation after the physical design is complete.

6.2 Implications and Future Research

This section examines the implications of this work and identifies some future research issues. The discussion focuses on the areas of robotics, system design, and design automation (DA) efforts.

Robotics

This work has demonstrated that ASIC technology can deliver superior performance needed for future robotic controllers. Effective utilization of this nascent technology calls for an algorithm-specific design approach. Functions in the device interaction layer of the robotic computation hierarchy are particularly suitable for ASIC implementation. With these functions executed neatly within individual ASIC chips, transfer of intermediate results are completely eliminated. This relieves the host system from not only the computation load, but also the communication burden. Thus, the host system can dedicate more resources to higher level tasks.

Since many robotic control functions share similar algorithmic characteristics with the IKS algorithm, namely, large granularity, strong serial data dependency, and the use of trigonometric functions, the idea of an on-chip cordic core is particularly attractive. The MACC architecture itself might be modified for executing other control functions. Hence, prototyping of the IKS chip is of practical interest. But before further work on physical design of the MACC chip proceeds, certain improvements at the architecture/logic level can be made. In the IKS algorithm, the configurations of the robot manipulator (left/right arm, up/down elbow, and flip/noflip wrist) can be specified by the user. Because of the format of the angle representation, this feature can be incorporated into the post-processing stage of the relevant cordic operations and efficiently implemented in hardware. Also, in this design, the algorithm implemented in the cordic core is based on the original cordic algorithm. More advanced techniques may be adopted to reduce the cordic iteration cycles. The original algorithm also has some convergence problems for the mode one operation, which can result in large computation errors when the robot is close to singularity positions. While the singularity problem is inherent in the robot arm design, the computational sensitivity to it can nonetheless be reduced. In the IKS algorithm, the mode one cordic operation is used only twice and the simulations uncover that the convergence problem actually occurs only in one particular operation, which also involves a constant. Thus, it appears that this problem can be solved with some special hardware tailored to this characteristic without resorting to more sophisticated, and usually more expensive, solutions.

System Design and Design Automation

The era of the microprocessor began in 1971 with Intel's introduction of the 4004. Since then, system design has matured into a discipline of building systems from chips. Before 1980 the semicustom design approach and all the basic ingredients of ASIC had already appeared, but the term ASIC became popular only after 1985. Interestingly, unlike the previous progress which lead to VLSI, it is difficult to single out a particular technological innovation, such as a new fabrication process, a memory chip, or a new microprocessor, that can symbolize the rise of ASIC. And yet this event is effecting a directional change in the discipline of system design. In retrospect, the force at work here appears to be the application-specific orientation which unites various technological advance-ments into a new direction. The impact is the increasingly stronger interactions among the three closely-related domains of technology, architecture, and application algorithms. A better understanding of these interactions thus becomes the key to meeting the ASIC challenge.

As more and more devices are fabricated on a single chip and powerful CAE tools for performing physical design tasks such as placement and routing become available, a new class of application-specific processors will emerge. To cope with the additional complexity created by this new situation, the time-proven strategy of ascending the abstraction ladder appears to be effective and architecture design has become a convenient focal point. But as one climbs up the abstraction ladder further and eventually reaches the application domain, the major performance enhancement opportunity may rely more on the designer's ability to take advantage of the application algorithm's characteristics than on the capability of a particular tool to minimize the number of gates or the routing area. The development of tools suitable for this new design orientation will become the major challenge for the years ahead.

It is against this background that the knowledge of the designer/engineer has been emphasized as complementary to the power of CAD tools throughout this book. The developments of the conceptual framework, the ASIC architecture design methodology, and the IKS chip design paradigm all reflect this orientation. By studying the IKS chip design paradigm, system designers can gain new insights on application-specific processor designs.

In addition to its educational values, the richness of design scenarios embodied in the paradigm also provide indispensable feedback for exploring ideas on new DA tools. Specifically, under the conceptual framework and the proposed ASIC architecture design methodology, design activities are viewed as making decisions regarding the manipulations of design information. By representing different facets of the design information in the forms of functional unit profile, dataflow table, and control signal pattern profile, many manipulation steps can and should be automated. Furthermore, the new concept of DISC design has been introduced to improve communication between designers and test engineers, and to facilitate maintenance and redesign/reuse efforts.

One particularly active R&D area relates to the development of VHDL. This work on VHDL has clearly shown the great promise of the language. Judging from the need for tools that are capable of supporting high-level abstractions, VHDL will enjoy wide acceptance in the next few years as it provides a vital link between expressions of very high-level application algorithms and low-level specifications of circiuts. Ultimately, this will change the infrastructure of electronic design by providing a new behavioral modeling layer evolving above the current circuit modeling layer based on SPICE.

Appendices

Appendix A.

The Closed Form IKS Algorithm for the Puma

$$r^2 = p_x^2 + p_y^2 \tag{A.1}$$

$$\theta_1 = \arctan\left[\frac{p_y}{p_x}\right] - \arctan\left[\frac{d_3}{\sqrt{r^2 - d_3^2}}\right] \tag{A.2}$$

$$f_{11p} = p_x C_1 + p_y S_1 \tag{A.3}$$

$$f_{11o} = o_x C_1 + o_y S_1 \tag{A.4}$$

$$f_{13o} = -o_x S_1 + o_y C_1 \tag{A.5}$$

$$f_{11a} = a_x C_1 + a_y S_1 \tag{A.6}$$

$$f_{13a} = -a_x S_1 + a_y C_1 \tag{A.7}$$

$$d = f_{11p}^2 + p_z^2 + (-d_4^2 - a_3^2 - a_2^2) \tag{A.8}$$

$$\theta_3 = \arctan\left[\frac{a_3}{-d_4}\right] - \arctan\left[\frac{d}{\sqrt{e^2 - d^2}}\right] \tag{A.9}$$

$$\text{where} \quad e = 2a_2\sqrt{a_3^2 + d_4^2}$$

$$w_1 = C_3 + (a_3/a_2) \tag{A.10a}$$

$$w_2 = S_3 + (d_4/a_2) \tag{A.10b}$$

$$\theta_{23} = \arctan\left[\frac{w_2 f_{11p} - w_1 p_z}{w_1 f_{11p} + w_2 p_z}\right] \tag{A.11}$$

$$\theta_2 = \theta_{23} - \theta_3 \tag{A.12}$$

$$\theta_4 = \arctan\left[\frac{f_{13a}}{C_{23}f_{11a} - S_{23}a_z}\right] \tag{A.13}$$

$$\theta_5 = \arctan\left[\frac{C_4(C_{23}f_{11a} - S_{23}a_z) + S_4 f_{13a}}{S_{23}f_{11a} + C_{23}a_z}\right] \tag{A.14}$$

$$\theta_6 = \arctan\left[\frac{-C_5[C_4(C_{23}f_{11o} - S_{23}o_z) + S_4 f_{13o}] + S_5(S_{23}f_{11o} + C_{23}o_z)}{-S_4(C_{23}f_{11o} - S_{23}o_z) + C_4 f_{13o}}\right] \tag{A.15}$$

Appendix B.

The IKS Algorithm in Pseudocodes

Task 1 (A.1):   cordic (3, px, py, x), R0 := Z;   /* output format: X_p, Z_a */

Task 2(A.2):   cordic (1, X, d3, x);   /* X: X_p */

Task 3 (A.2):   cordic (3, X, d3, x);   /* X: X_p, Z: Z_a */

Task 4 (A.2):   J9 := R0 - Z;   /* J9=q1 */

Task 5 (A.3):   cordic (2, K1, x, Z), R0 := X, R1 := Y; /* X: X_o, Y: Y_o */

Task 6 (A.3):   R2 := mac (R0, px, +, R1, py);

Task 7 (A.8):   R3 := R2 - d4, R4 := R2 + d4;

Task 8 (A.8):   Z := pz - a2, X:= px + a2;

Task 9 (A.8):   R3 := mac (Z, X, +, R3, R4) - K2   /* K2 = a3*a3, Z: Z_s */

Task 10 (A.9):   cordic (1, e, Z, x);   /* X: X_s */

Task 11 (A.9):   cordic (3, X, R3, x);   /* Z: Z_a */

Task 12 (A.9):   J11 := K3 - Z;   /* J11 = q3 */

Task 13 (A.10):   cordic (2, K1, x, J11);   /* X: X_o, Y: Y_o */

Task 14 (A.10a): R4 := X + K4;

Task 15 (A.11):   R3 := mac (R5, R2, -, R4, pz) + R2;

Task 16 (A.11):   Z := mac (R4, R2, +, R5, pz) + pz;

Task 17 (A.11):   cordic (3, Z, R3, x), J15 := Z;   /* Z: Z_a */

Task 18 (A.13):   cordic (2, K1, x, Z), R3 := X, R4 := Y; /* X: X_o, Y: Y_o */

Task 19 (A.12):   J10 := J15 - J11;   /* J10 = q2 */

Task 20 (A.6):   R6 := mac (R0, ax, +, R1, ay);

Task 21 (A.7):   R5 := mac (R0, ay, -, R1, ax);

Task 22 (A.4):   R2 := mac (R0, ox, +, R1, oy);

Task 23 (A.5):   R0 := mac (R0, oy, -, R1, ox);

Task 24 (A.13):   R7 := mac (R3, az, +, R4, R6);

Task 25 (A.13):   R1 := mac (R3, R6, -, R4, az);

Task 26 (A.15):   R6 := mac (R3, R2, -, R4, oz);

Task 27 (A.15):   R4 := mac (R3, oz, +, R4, R2);

Task 28 (A.13):   cordic (3, R1, R5, x), J12 := Z;   /* Z: Z_a, J12 = q4 */

Task 29 (A.14):   cordic (2, K1, x, Z), R2 := X, R3 := Y; /* X: X_o, Y: Y_o */

Task 30 (A.14):   R1 := mac (R2, R1, +, R3, R5);

Task 31 (A.15):   R5 := mac (R2, R6, +, R3, R0);

Task 32 (A.15):   R6 := mac (R2, R0, -, R3, R6);

Task 33 (A.14):   cordic (3, R7, R1, x), J13 := Z;   /* Z: Z_a, J13 = q5 */

Task 34 (A.15):   cordic (2, K1, x, Z), R2 := X, R3 := Y; /* X: X_o, Y: Y_o */

Task 35 (A.15):   Z := mac (R3, R4, -, R2, R5);

Task 36 (A.15):   cordic (3, R6, Z, x), J14 := Z;   /* Z: Z_a, J14 = q6 */

Appendix C.

Control Signal Definition

1. Transmission Gate (TG[8:0])

Signal Name	Index	Source	Destination
R1-A	[8,7]	00: Disable	–
		01: Z-Adder	R Reg. Port 1
		10: R Reg. Port 1	Bus A
		11: Bus A	R Reg. Port 1
R2_B	[6]	R Reg. Port 2	Bus B
J_A	[5]	J Reg.	Bus A
A_J	[4]	Bus A	J Reg.
M_B	[3]	ROM M	Bus B
X_A	[2]	X Adder output	Bus A
Z_A	[1]	Z Adder output	Bus A
P_A	[0]	MPYer output	

2. Adder Operand Control

Signal	Bits	Definition
x1	1	0: x1 = 0
		1: x1 = X
x2	1	0: x2 = R-shift (Y, i)
		1: x2 = Bus B
y1	1	0: y1 = 0
		1: y1 = Y
y3	2	00: y2 = 0
		01: y2 = Bus A if f = 0
		= 0 if f = 1
		10: y2 = Bus A
		11: y2 = R-shift (X, i)
z1	2	00: z1 = 0
		01: z1 = P
		10: z1 = Bus A
z2	2	00: z2 = 0
		01: x2 = Bus B
		10: z2 = Z_ang

3. Arithmetic/Logic Operation Control

Signal Name	Bits	Category	Operation
Z	2	iv	$Z_0 = 0$: $Z_c = Z_1$ $Z_0 = 1$: $Z_c = A_z$
XY	2	iv	00: $X_c = Y_c = 0$ 01: $X_c = A_x$, $Y_c = A_y$ 10: $X_c = Y_c = f$ 11: $X_c = Y_c = 1$
setf	2	iii	00: f unchange 01: $f = y2_{n-1}$ 10: $f = Z_{n-1}$ 11: $f = Z_{n-1} \oplus Z_{n-2}$
m	2	ii	00: non-cordic mode 01: cordic – root difference square 10: cordic – sin/cos 11: dcordic – arctan
sft_i	5	ii	right shift X and Y *i* bits
zang	2	ii	0X: $z2 = Z$ 10: $z2 = Z$ with $Z_{n-1} = f \oplus Z_{n-1}$ 11: $z2 = Z$ with $Z_{n-1} = (Z_{n-1} \oplus Z_{n-2}) \oplus Z_{n-1}$

Appendix D.

The MACC Encoding Scheme and Code-Maps

1. Signals for Setting up the Data Routes — (a) Encoding Scheme

Phase 1 Pattern							Phase 2 Pattern						Encoded as
Bus src	A dst	Bus src	B dst	Reg J	Reg R	zang	z1	z2	x1	x2	y1	y2	Code
						0	0	0	0	Y	0	X	0
Z	z1	M	z2			0	A	B	X	Y	Y	X	1
					w,Z		P	0	0	Y	Y	0	2
J	y2			r		0	0	0	0	Y	0	A	3
J	z1			r		0	A	Z	0	Y	Y	0	4
Z	y2					0	P	0	0	Y	0	A	5
Z	y2					0	0	0	0	Y	0	A	6
R	y2				r	0	P	Z	0	Y	0	A	7
R	y2				r	0	0	0	0	Y	0	A	8
R	y2				r	0	P	Z	0	Y	Y	A	9
J	R			r	w,A		P	0	0	Y	0	X	10
X	R	R	z2		r,w,A		0	B	0	Y	0	0	11
J	m1	R	m2	r		0	P	0	0	Y	0	0	12
J	m1	R	m2	r		0	0	Z	0	Y	0	0	13
J	m1	R	m2	r		0	P	Z	X	Y	0	0	14
J	m1	R	m2	r	w,Z	0	P	0	0	Y	0	0	15
R	m1	R	m2		r	0	P	Z	0	Y	0	0	16
R	m1	R	m2		r	0	P	0	X	Y	0	0	17
R	m1	R	m2		r	0	0	Z	0	Y	0	0	18
X	m1	M	m2			1	0	ang	0	Y	0	0	19
Z	z1	M	z2		w,Z	0	A	B	0	Y	0	0	20
R	z1	M	z2		r	0	A	B	0	Y	0	0	21
J	z1	M	z2	r	w,Z	0	A	B	0	Y	0	0	22
R	z1	R	z2		r	0	A	B	0	Y	0	0	23
R	y2	M	x2		r	0	0	Z	0	B	0	A	24
P	y2	M	x2			0	0	0	0	B	0	A/0	25
X	R	M	x2		w,A		0	0	0	B	Y	0	26
Z	J	M	x2	w		3	0	ang	0	B	0	0	27
Z	J	M	x2	w		1	0	ang	0	B	0	0	28
						0	0	Z	0	Y	0	X	29

1. Signals for Setting up the Data Routes — (b) Code-Map

	DECODED SIGNAL VALUE																		
	ctrl_v1 (0-10)												ctrl_v2 (0-8)						
	tg [9]									zang		x1	x2	y1	y2[2]		z1[2]		z2[2]	
	[8]	[7]	[6]	[5]	[4]	[3]	[2]	[1]	[0]	[1] [0]					[1]	[0]	[1]	[0]	[1]	[0]
Opcode 1 no.	R1A1	R1A0	R2|B	J|AA	A|JJ	M|BB	X|AA	Z|AA	P|AA											
0	0	0	0	0	0	0	0	0	0	0 0		0	0	0	1	1	0	0	0	0
1	0	0	0	0	0	1	0	1	0	0 0		1	0	1	1	1	1	0	0	1
2	0	1	0	0	0	0	0	0	0	0 0		0	0	1	0	0	0	1	0	0
3	0	0	0	1	0	0	0	0	0	0 0		0	0	0	1	0	0	0	0	0
4	0	0	0	1	0	0	0	0	0	0 0		0	0	1	0	0	1	0	1	0
5	0	0	0	0	0	0	0	1	0	0 0		0	0	0	1	0	0	1	0	0
6	0	0	0	0	0	0	0	1	0	0 0		0	0	0	1	0	0	0	0	0
7	1	0	0	0	0	0	0	0	0	0 0		0	0	0	1	0	0	1	1	0
8	1	0	0	0	0	0	0	0	0	0 0		0	0	0	1	0	0	0	0	0
9	1	0	0	0	0	0	0	0	0	0 0		0	0	1	1	0	0	1	1	0
10	1	1	0	1	0	0	0	0	0	0 0		0	0	0	1	1	0	1	0	0
11	1	1	1	0	0	0	1	0	0	0 0		0	0	0	0	0	0	0	0	1
12	0	0	1	1	0	0	0	0	0	0 0		0	0	0	0	0	0	1	0	0
13	0	0	1	1	0	0	0	0	0	0 0		0	0	0	0	0	0	0	1	0
14	0	0	1	1	0	0	0	0	0	0 0		1	0	0	0	0	0	1	1	0
15	0	1	1	1	0	0	0	0	0	0 0		0	0	0	0	0	0	1	0	0
16	1	0	1	0	0	0	0	0	0	0 0		0	0	0	0	0	0	1	1	0
17	1	0	1	0	0	0	0	0	0	0 0		1	0	0	0	0	0	1	0	0
18	1	0	1	0	0	0	0	0	0	0 0		0	0	0	0	0	0	0	1	0
19	0	0	0	0	0	1	1	0	0	0 1		0	0	0	0	0	0	0	1	0
20	0	1	0	0	0	1	0	1	0	0 0		0	0	0	0	0	1	0	0	1
21	1	0	0	0	0	1	0	0	0	0 0		0	0	0	0	0	1	0	0	1
22	0	1	0	1	0	1	0	0	0	0 0		0	0	0	0	0	1	0	0	1
23	1	0	1	0	0	0	0	0	0	0 0		0	0	0	0	0	1	0	0	1
24	1	0	0	0	0	1	0	0	0	0 0		0	1	0	1	0	0	0	1	0
25	0	0	0	0	0	1	0	0	1	0 0		0	1	0	0	1	0	0	0	0
26	1	1	0	0	0	1	1	0	0	0 0		0	1	1	0	0	0	0	0	0
27	0	0	0	1	1	1	0	1	0	1 1		0	1	0	0	0	0	0	1	0
28	0	0	0	1	1	1	0	1	0	0 1		0	1	0	0	0	0	0	1	0
29	0	0	0	0	0	0	0	0	0	0 0		0	0	0	1	1	0	0	1	0

2. Signals for Controlling Operations — (a) Encoding Scheme

Adders Z	X	Y	self	Encoded as
+	+	+	0	0
from C.C.*			0	1
−	+	+	0	2
+	f	f	0	3
+	+	+	1	4
−	+	+	1	5
+	+	+	2	6
+	+	+	3	7

* From the Cordic Control unit.

2. Signals for Controlling Operations — (b) Code-Map

Opcode2	Z [1] [0]	XY [1] [0]	self [1] [0]
0	0 0	0 0	0 0
1	0 1	0 1	0 0
2	1 0	0 0	0 0
3	0 0	1 0	0 0
4	0 0	0 0	0 1
5	1 0	0 0	0 1
6	0 0	0 0	1 0
7	0 0	0 0	1 1

Appendix E.

The MACC Microcode for Computing the IKS

Store Addr. [7]	Clock Cycle	Opcode 1 [5]	Opcode 2 [3]	m [2]	Parameters J [4]	R1 [3]	M/R2 [5]
0	0	0	0	0			
1	1	3	4	0	6		
2	2	3	3	0	7		
3	3	1	1	3			
4	27	19	0	0			24
5	28	2	0	0		0	
6	29	17	0	0		[ ]	[ ]
7	30	20	2	0		7	30
8	31	24	0	0		7	30
9	32	29	6	0			
10	33	1	1	1			
11	56	19	0	0			25
12	57	10	0	0	8	7	
13	58	25	4	0			30
14	59	0	3	0			
15	60	1	1	3			
16	84	19	0	0			[ ]
17	85	2	0	0		6	
18	86	23	2	0		0	6
19	87	27	7	0	9		24
20	88	1	1	2			
21	112	0	3	0			
22	113	11	0	0		1	[ ]
23	114	11	2	0		0	7
24	115	15	0	0	7	7	1
25	116	12	0	0	6		0
26	117	5	0	0			
27	118	7	0	0			
28	119	20	2	0		2	31
29	120	2	0	0		3	
30	121	21	0	0		2	31
31	122	22	2	0	8	4	31
32	123	22	0	0	8	5	31
33	124	2	0	0		6	
34	125	17	0	0		3	4
35	126	17	0	0		5	6
36	127	12	0	0	[ ]		[ ]
37	128	7	0	0			
38	129	1	2	0			26
39	130	20	2	0		3	27
40	131	18	2	0			
41	132	24	6	0		3	27

Addr.	Clock	Opcode 1	Opcode 2	m	J	R1	M/R2
42	133	1	1	1			
43	156	19	0	0			25
44	157	0	0	0			
45	158	25	4	0			[ ]
46	159	8	3	0		3	
47	160	1	1	3			
48	184	19	2	0			
49	185	1	0	0			28
50	186	27	7	0	11		24
51	187	1	1	2			
52	211	0	3	0			
53	212	26	0	0		5	29
54	213	17	0	0	0	5	2
55	214	11	0	0		4	[ ]
56	215	17	0	0		4	7
57	216	13	0	0	8		5
58	217	16	0	0		4	2
59	218	5	0	0			
60	219	9	0	0		2	
61	220	4	0	0	8		
62	221	6	4	0			
63	222	0	3	0			
64	223	1	1	3			
65	247	19	0	0			[ ]
66	248	28	0	0	15		24
67	249	1	1	2			
68	273	0	3	0			
69	274	11	0	0		4	[ ]
70	275	11	0	0		3	[ ]
71	276	13	0	0	3		0
72	277	13	0	0	4		1
73	278	12	0	0	3		1
74	279	14	0	0	4		0
75	280	15	0	0	0	6	0
76	281	14	2	0	1		1
77	282	15	0	0	0	5	1
78	283	14	0	0	1		0
79	284	15	0	0	5	2	3
80	285	16	2	0		4	6
81	286	15	0	0	5	0	4
82	287	16	0	0		3	6
83	288	15	0	0	2	7	4
84	289	16	2	0		3	2

Addr.	Clock	Opcode 1	Opcode 2	m	J	R1	M/R2
85	290	15	0	0	2	1	3
86	291	16	2	0		4	2
87	292	2	0	0		6	
88	293	7	4	0		1	
89	294	2	0	0		4	
90	295	8	3	0		5	
91	296	1	1	3			
92	320	19	0	0			[ ]
93	321	28	0	0	12		24
94	322	1	1	2			
95	346	0	3	0			
96	347	11	0	0		3	[ ]
97	348	11	0	0		2	[ ]
98	349	18	0	0		2	1
99	350	18	0	0		3	5
100	351	17	0	0		2	6
101	352	16	0	0		3	0
102	353	2	0	0		1	
103	354	18	0	0		3	6
104	355	16	0	0		2	0
105	356	2	0	0		5	
106	357	7	5	0		7	
107	358	2	0	0		6	
108	359	8	3	0		1	
109	360	1	1	3			
110	384	19	0	0			[ ]
111	385	28	0	0	13		24
112	386	1	1	2			
113	410	10	3	0	15	2	
114	411	11	0	0		3	[ ]
115	412	11	0	0		1	[ ]
116	413	16	2	0		1	5
117	414	18	0	0		3	4
118	415	10	3	0	11	0	
119	416	7	5	0		6	
120	417	6	3	0			
121	418	1	1	3			
122	442	19	0	0			[ ]
123	443	27	7	0	14		[ ]
124	444	23	2	0		2	0
125	445	27	7	0	10		[ ]

Bibliography

[Ada86]   J. A. Adam, "Technology '86: aerospace and military," *IEEE Spectrum*, vol. 23, pp. 76–81, Jan. 1986.

[Ale*t*83]   J. S. Albus, *et al.*, "Hierarchical Control for Robots in an Automated Factory," *Proc. 13th Int. Symp. Ind. Robots*, pp. 13-29 to 13-43, 1983.

[Amd67]   G. M. Amdahl, "Validity of the single processor approach to achieving large scale computing capabilities," *Proc. AFIPS.*, vol. 30, Thompson, Washington D.C., 1967, pp. 483–485.

[Ano87]   Anonymous, "NCR-MCC Design Advisor," *IEEE Design Test Comput.*, p. 67, Aug. 1987.

[Arm89]   J. Armstrong, *Chip Level Modeling with VHDL*. Englewood Cliffs, NJ: Prentice Hall, 1989.

[Ary85]   S. Arya, "An optimal instruction-scheduling model for a class of vector processors," *IEEE Trans. Comput.*, vol. C-34, pp. 981–995, Nov. 1985.

[AtSe88]   W. C. Athas and C. L. Seitz, "Multicomputers: message-passing concurrent computers," *Computer*, vol. 21, pp. 9–24, Aug. 1988.

[BeNe71]   C. G. Bell and A. Newell, *Computer Structures: Readings and Examples*. New York, NY: McGraw Hill, 1971.

[Ber85]   R. Beresford, "A profile of current applications of gate arrays and standard-cell ICs," *VLSI Syst. Des.*, pp. 62–66, Sept. 1985.

[BeSz87]   A. K. Bejczy and Z. Szakaly, "Universal computer control system (UCCS) for space telerobots," *Proc. IEEE 1987 Int. Conf. Robotics Automation*, pp. 318–324, 1987.

[BiMR83]   R. Bisiani, H. Mauersberg, and R. Reddy, "Task-oriented architectures," *Proc. IEEE*, vol. 71, pp. 885–898, July 1983.

[BPTP87] B. K. Bose, L. Pei, G. S. Taylor, and D. A. Patterson, "Fast multiply and divide for a VLSI floating-point unit," *Proc. IEEE 8th Symp. Comput. Arith.*, pp. 87–94, 1987.

[BrGr87] J. Brouwers and M. Gray, "Integrating the electronic design process," *VLSI Syst. Des.*, pp. 38–47, June 1987.

[BuMa85] M. R. Buric and T. G. Matheson, "Silicon compilation environments," *Proc. IEEE 1985 Custom IC Conf.*, pp. 208–212, 1985.

[CaLu87] J. R. Cavallaro and F. T. Luk, "Cordic arithmetic for an SVD processor," *Proc. IEEE 8th Symp. Comput. Arith.*, pp. 113–120, 1987.

[CFAB86] J. B. Chen, R. S. Fearing, B. S. Armstrong, and J. W. Burdick, "NYMPH: A multiprocessor for manipulation applications," *Proc. IEEE 1986 Int. Conf. Robotics Automation*, pp. 1731–1736, 1986.

[Che86] J. Y. Chen, "CMOS — the emerging VLSI technology," *IEEE Circ. Dev. Mag.*, pp. 16–31, Mar. 1986.

[Coe89] D. R. Coelho, *VHDL Hardware Design Techniques*. Norwell, MA: Kluwer Academic, 1989.

[Coh85] P. A. Cohen, "Whatever happened to the robot boom?" *Proc. Robots/9*, pp. 13-22 to 13-32, 1985.

[Col88a] B. C. Cole, "Intel's ambitious game plan in embedded chips," *Electronics*, pp. 97–100, Apr. 14, 1988.

[Col88b] B. C. Cole, "RISC slugfest: Is marketing more important than performance?" *Electronics*, pp. 63–68, Apr. 28, 1988.

[Cra86] J. J. Crag, *Introduction to Robotics: Mechanics and Control*. Reading, MA: Addison-Wesley, 1986.

[Cus88a] B. Cushman, "RISC changes the balance," *VLSI Syst. Des.*, pp. 64–69, June 1988.

[Cus88b] B. Cushman, "Surveying the RISC realm," VLSI Syst. Des., pp. 64–63, 100, July 1988.

[DaDo86] W. K. Dawson and R. W. Dobinson, "A framework for computer design," *IEEE Spectrum*, vol. 23, pp. 49–54, Oct. 1986.

[Das84] S. Dasgupta, *The Design and Description of Computer Architectures*. New York, NY: Wiley, 1984, pp. 1–39.

[Des84] A. M. Despain, "Notes on computer architecture for high performance," in *New Computer Architectures*, J. Tiberghien, Ed. New York, NY: Academic Press, 1984, pp. 81–98.

[D&T86] Special issue on VHDL: The VHSIC Hardware Description Language, *IEEE Design Test Comput.*, vol 3, no.2, Apr. 1986.

[EBCH86] C. K. Erdelyi, R. A. Bechade, M. P. Concannon, and W. K. Hoffman, "Custom and semicustom design," in *Design Methodologies*, S. Goto, Ed. New York, NY: Elsevier, 1986, ch. 1.

[Far81] P. M. Farmwald, "High bandwidth evaluation of elementary functions," *Proc. IEEE 5th Symp. Comput. Arith.*, pp. 139–142, 1981.

[FePa86] C. F. Fey and D. Paraskevopoulos, "A comparison of product costs using MSI, gate arrays, standard cells, and full custom VLSI," *IEEE J. Solid-State Circ.*, vol. SC-21, pp. 297–303, Apr. 1986.

[FePa87] C. F. Fey and D. Paraskevopoulos, "A techno-economic assessment of application-specific integrated circuits: current status and future trends," *Proc. IEEE*, vol. 75, pp. 829–841, June 1987.

[Fer85] D. K. Ferry, "Interconnection lengths and VLSI," *IEEE Circ. Dev. Mag.*, pp. 39–42, July 1985.

[FiFi87] M. A. Fischler and O. Firschein, \fIIntelligence: The Eye, the Brain, and the Computer. Reading, MA: Addison-Wesley, 1987, ch. 7.

[GaKu83] D. D. Gajski and R. H. Kuhn, "New VLSI tools," *Computer*, vol. 16, pp. 11–14, Dec. 1983.

[GaPe85] D. D. Gajski and J.-K. Peir, "Essential issues in multiprocessor systems," *Computer*, vol. 18, pp. 9–27, June, 1985.

[GaRo84] D. B. Gannon and J. V. Rosendale, "On the impact of communication complexity on the design of parallel numerical algorithms," *IEEE Trans. Comput.*, vol. C-33, pp. 1180–1194, Dec. 1984.

[GFCM87] D. Gauthier, P. Freedman, G. Carayannis, and A. S. Malowany, "Interprocess communication for distributed robotics," *IEEE J. Robotics Automation*, vol. RA-3, pp. 493–504, Dec. 1987.

[GiMi87] C. E. Gimarc and V. M. Milutinovic, "A survey of RISC processors and computers of the mid-1980s," *Computer*, vol. 20, pp. 59–69, Sept. 1987.

[Gol84] S. M. Goldwasser, "Computer architecture for grasping," *Proc. IEEE 1984 Int. Conf. Robotics Automation*, pp. 320–325, 1984.

[GoSt87] Y. Goto and A. Stentz,"The CMU system for mobile robot navigation," *Proc. IEEE 1987 Int. Conf. Robotics Automation*, pp. 99–105, 1987.

[GuKa85] K. C. Gupta and K. Kazerounian, "Improved numerical solutions of inverse kinematics of robots," *Proc. IEEE 1985 Int. Conf. Robotics Automation*, pp. 743–748, 1985.

[Har82] L. D. Harmon, "Automated tactile sensing," *Int. J. Robotics Res.*, vol. 1, pp. 3–32, Summer 1982.

[Har86] A. C. Hartmann, "Software or silicon? The designer's option," *Proc. IEEE*, vol. 74, pp. 861–874, June 1986.

[HaTu80] G. L. Haviland and A. A. Tuszynski, "A cordic arithmetic processor chip," *IEEE Trans. Comput.*, vol. C-29, pp. 68–79, Feb. 1980.

[HeHk87]   J. N. Herndon, W. R. Hamel and D. P. Kuban, "Traction-drive telerobot for space manipulation," *Proc. IEEE 1987 Int. Conf. Robotics and Automation*, pp. 450–455, 1987.

[Hen84]   J. L. Hennessy, "VLSI processor architecture," *IEEE Trans. Comput.*, vol. C-33, pp. 1221–1246, Dec. 1984.

[Hin88]   J. Hinkle, private communication, 1988.

[HiRa87]   V. L. Hinder and A. S. Rappaport, "Employing semicustom: A study of users and potential users," *VLSI Syst. Des.*, pp. 6–25, May 20, 1987.

[Hna87]   E. R. Hnatek, *Integrated Circuit Quality and Reliability*. New York, NY: Marcel Dekker, 1987, pp. 271–298.

[HnWi85]   E. R. Hnatek and B. R. Wilson, "Practical considerations in testing semicustom and custom ICs," *VLSI Des.*, pp. 20–42, Mar. 1985.

[Hoet87]   R. Hornung, *et al.*, "A versatile VLSI design system for combining gate array and standard cell circuits on the same chip," *Proc. IEEE 1987 Custom IC Conf.*, pp. 245–247, 1987.

[Hol87]   E. E. Hollis, *Design of VLSI Gate Array ICs*. Englewood Cliffs, NJ: Prentice-Hall, 1987, pp. 207–218.

[HoRe77]   R. W. Hon and D. R. Reddy, "The effect of computer architecture on algorithm decomposition and performance," in *High Speed Computer and Algorithm Organization*, D. J. Kuck, *et al.*, Ed. New York, NY: Academic Press, 1977, pp. 411–423.

[Hui88]   L. M. Huisman, "The reliability of approximate testability measures," *IEEE Design Test Comput.*, pp. 57–67, Dec. 1988.

[Hur85]   S. L. Hurst, *Custom-Specific Integrated Circuits, Design and Fabrication*. New York, NY: Marcel Dekker, 1985, ch. 1.

[HwBr84]   K. Hwang and F. A. Briggs, *Computer Architecture and Parallel Processing*. New York, NY: McGraw Hill, 1984, ch. 2.

[HwWX87]   K. Hwang, H. C. Wang, and Z. Xu, "Evaluating elementary functions with Chebyshev Polynomials on pipeline nets," *Proc. IEEE 8th Symp. Comput.* Arith., pp. 121–128, 1987.

[Jac86]   H. Jacobs, "Verification of a second-generation 32-bit microprocessor," *Computer*, vol. 19, pp. 64–70, Apr. 1986.

[Jaet85]   H. V. Jagadish, *et al.*, "A study of pipelining in computing arrays," *IEEE Trans. Comput.*, vol. C-35, pp. 431–440, May 1985.

[JaGD87]   L. H. Jamieson, D. B. Gannon, and R. J. Douglass, Eds., *The Characteristics of Parallel Algorithms*. Cambridge, MA: MIT Press, 1987.

[Jam87]   L. H. Jamieson, "Characterizing parallel algorithms," in *The Characteristics of Parallel Algorithms*, L. H. Jamieson, D. B. Gannon, and R. J. Douglass, Eds. Cambridge, MA: MIT Press, 1987, pp. 65–100.

[JaOr87]   M. Amin-Javaheri and D. E. Orin, "A systolic architecture for computation of the manipulator inertia matrix," *Proc. IEEE 1987 Int. Conf. Robotics Automation*, pp. 647–653, 1987.

[Jar84]   J. F. Jarvis, "Robotics," *Computer*, vol. 17, pp. 283–292, Oct. 1984.

[KaNa85]   H. Kasahara and S. Narita, "Parallel processing of robot-arm control computation on a multimicroprocessor system," *IEEE J. Robotics Automation*, vol. RA-1, pp. 104–113, June 1985.

[Kat85]   R. H. Katz, "Computer-aided design databases," *IEEE Design Test Comput.*, pp. 70–74, Feb. 1985.

[KaWo85]   W. C. Kabat and A. S. Wojcik, "Automated synthesis of combinational logic using theorem-proving techniques," *IEEE Trans. Comput.*, vol. C-34, pp. 610–632, July 1985.

[Kee86]   K. Keenan, "A third generation robot controller," *Ind. Robot, 16th ISIR/Robotex*, pp. 9–11, Oct. 1986.

[Kle85]   L. Kleinrock, "Distributed systems," *Computer*, vol. 18, pp. 90–103, Nov. 1985.

[KlWa82]   C. A. Klein and W. Wahawisan, "Use of a multiprocessor for control of a robotic system," *Int. J. Robotics Res.*, vol. 1, pp. 45–59, Summer 1982.

[Kuc77]   D. J. Kuck, "A survey of parallel machine organization and programming," *Comput. Survey*, vol. 9, pp. 29–59, Mar. 1977.

[Kun80]   H. T. Kung, "The structure of parallel algorithms," in *Advance in Computers*, vol. 19. New York, NY: Academic Press, 1980, pp. 65–111.

[Kun84]   H. T. Kung, "Putting inner loops automatically in silicon," in *VLSI Engineering Beyond Software Engineering*, T. Kunii, Ed. New York, NY: Springer-Verlag, 1984, pp. 70–104.

[LaRu71]   B. S. Landman and R. L. Russo, "On a pin vs. block relationship for partitions of logic graph," *IEEE Trans. Comput.*, vol. 20, pp. 1469–1479, Dec. 1971.

[LeCh87]   C. S. G. Lee and P. R. Chang, "A maximum pipelined CORDIC architecture for inverse kinematic position computation," *IEEE J. Robotics Automation*, vol. RA-3, pp. 445–458, Oct. 1987.

[LeFS88]   S. S. Leung, P. D. Fisher, and M. A. Shanblatt, "A conceptual framework for ASIC design," *Proc. IEEE*, vol. 76, pp. 741–755, July 1988.

[LeLe84]   C. S. G. Lee and B. H. Lee, "Resolved motion adaptive control for mechanical manipulators," *Trans. ASME, J. Dynam. Syst., Meas. Contr.*, vol. 106, no. 2, pp. 134–142, June 1984.

[LeSh87]   S. S. Leung and M. A. Shanblatt, "Real-time DKS on a single chip," *IEEE J. Robotics Automation*, vol. RA-3, pp. 281–290, Aug. 1987.

[LeSh88a]   S. S. Leung and M. A. Shanblatt, "Computer architecture design for robotics," *Proc. IEEE 1988 Int. Conf. Robotics Automation*, pp. 453–456, 1988.

[LeSh88b]   S. S. Leung and M. A. Shanblatt, "Designing algorithm-specific computational hardware for robotic control," *Proc. 2nd Int. Symp. Robotics Manufact. Res., Edu. and Appl.*, pp. 109–118, 1988.

[Leu89a]   S. S. Leung, *A Design Paradigm for Implementing Robotic Control Algorithms in ASIC*. Ph.D. Diss., Michigan State University, Feb. 1989.

[Leu89b]   S. S. Leung, "Behavioral modeling of transmission gates in VHDL," *Proc. 26th IEEE/ACM DAC*, 1989.

[Lin87]   J. R. Lineback, "LSI logic's giant array breaks the record for usable gates," *Electronics*, pp. 55–56, Oct. 29, 1987.

[Lip83]   H. M. Lipp, "Methodical Aspects of Logic Synthesis," *Proc. IEEE* vol. 71, pp. 88–97, Jan. 1983.

[LRM88]   *IEEE Standard VHDL Language Reference Manual*. New York, NY: IEEE, 1988.

[LSI87a]   *Databook: 1.5-Micron Compacted Array Technology*. Milpitas, CA: LSI Logic, 1987.

[LSI87b]   *Databook: HCMOS Megafunction*. Milpitas, CA: LSI Logic, 1987.

[LuWP80]   J. Y. S. Luh, M. W. Walker, and R. P. C. Paul, "On-line computation scheme for mechanical manipulators," *Trans. ASME, J. Dynam. Syst., Meas. Contr.*, vol. 120, pp. 69–76, June 1980.

[Man88]   T. Manuel, "Taking a close look at the Motorola 88000," *Electronics*, pp. 75–78, Apr. 28, 1988.

[May85]   E. D. Maynard Jr., "VHSIC database and database management system requirements," *VLSI Des.*, pp. 9–96, Feb. 1985.

[McC86]   E. J. McClusky, *Logic Design Principles*. Englewood Cliffs, NJ: Prentice-Hall, 1986, ch. 10.

[McD85]   K. J. McDermott, "A pragmatic state of the art analysis of industrial robot functions and characteristics,"*Proc. Robots/9*, pp. 20-67 to 20-77, 1985.

[McL86]   J. McLeod, "Computer aided design and engineering," Electronics, pp. 94-97, Oct. 16, 1986.

[McL89]   J. McLeod, "A giant leap for simulation," *Electronics*, pp. 73–76, Feb. 1989.

[Mey86]   E. L. Meyer, "Gate array testability: A customer perspective," *VLSI Syst. Des.*, pp. 34–42, June 1986.

[Mon86]   M. D. Montemerlo, "NASA's automation and robotics technology development program," *Proc. IEEE 1986 Int. Conf. Robotics Automation*, pp. 977–986, 1986.

[Nas85]   J. G. Nash, "A systolic/cellular computer architecture for linear algebraic operations," *Proc. IEEE 1985 Int. Conf. Robotics Automation*, pp. 779–784, 1985.

[NeVi86] A. R. Newton and A. L. Sangiovanni-Vincentelli, "Computer-aided design for VLSI circuits," *Computer*, pp. 38–60, Apr. 1986.

[NeVi87] A. R. Newton and A. L. Sangiovanni-Vincentelli, "CAD tools for ASIC design," *Proc. IEEE*, vol. 75, pp. 765–776, June 1987.

[NiHw85] L. M. Ni and K. Hwang, "Vector-reduction techniques for arithmetic pipelines," *IEEE Trans. Comput.*, vol. C-34, pp. 404–411, May 1985.

[NiLe85] R. Nigam and C. S. G. Lee, "A multiprocessor-based controller for the control of mechanical manipulators," *IEEE J. Robotics Automation*, vol. RA-1, pp. 173–182, Dec. 1985.

[NSHB86] S. Narasimhan, D. Siegel, J. M. Hollerbach, K. Biggers, and G. Gerpheide, "Implementation of control methodologies on the computational architecture for the Utah/MIT hand," *Proc. IEEE 1986 Int. Conf. Robotics Automation*, pp. 1884–1889, 1986.

[OCOS85] D. E. Orin, H. H. Chao, K. W. Olson, and W. W. Schrader, "Pipeline/parallel algorithms for the Jacobian and inverse dynamic computations," *Proc. IEEE 1985 Int. Conf. Robotics Automation*, pp. 785–789, 1985.

[OkSG86] N. Okuda, M. Sugai, and N. Goto, "Semicustom and custom LSI technology," *Proc. IEEE*, vol. 74, pp. 1636–1645, Dec. 1986.

[OrTs86] D. E. Orin and Y. T. Tsai, "A real-time computer architecture for inverse kinematics," *Proc. IIEEE 1986 Int. Conf. Robotics Automation*, pp. 843–850, 1986.

[PaAk83] C. A. Palesko and L. A. Akers, "Logic partitioning for minimizing gate arrays," *IEEE Trans. CAD IC Syst.*, vol. CAD-2, pp. 117–121, Apr. 1983.

[PaHa87] A. C. Parker and S. Hayati, "Automating the VLSI design process using expert systems and silicon compilation," *Proc. IEEE*, vol. 75, pp. 777–785, June 1987.

[PaKL80] D. A. Padua, D. J. Kuck, and D. H. Lawrie, "High-speed multiprocessors and compilation techniques," *IEEE Trans. Comput.*, vol. C-29, pp. 763–776, Sept. 1980.

[Pan86]   D. Pantic, "Benefits of integrated-circuit burn-in to obtain high reliability parts," *IEEE Trans. Reliab.*, vol. R-35, pp. 3–6, Apr. 1986.

[Pau81]   R. P. Paul, *Robot Manipulators: Mathematics, Programming, and Control*. Cambridge, MA: MIT Press, 1982.

Index